I0482716

NIST Technical Note 1453

Smoke Component Yields from Room-scale Fire Tests

Richard G. Gann
Jason D. Averill
Erik L. Johnsson
Marc R. Nyden
Richard D. Peacock

Fire Research Division
Building and Fire Research Laboratory
National Institute of Standards and Technology
Gaithersburg, MD 20899-8664

April 2003

U.S. Department of Commerce
Donald L. Evans, Secretary

Technology Administration
Phillip J. Bond, Under Secretary for Technology

National Institute of Standards and Technology
Dr. Arden L. Bement, Jr., Director

ABSTRACT

This report presents the methodology for and results from a series of room-scale fire tests to produce data on the yields of toxic products in both pre-flashover and post-flashover fires. The combustibles examined were: a sofa made of upholstered cushions on a steel frame, particleboard bookcases with a laminated finish, polyvinyl chloride sheet, and household electric cable. They were burned in a room with a long adjacent corridor. The yields of CO_2, CO, HCl, HCN, and carbonaceous soot were determined. Other toxicants (e.g., NO_2, formaldehyde and acrolein) were not found; concentrations below the detection limits were shown to be of limited toxicological importance relative to the detected toxicants. The toxicant yields from sofa cushion fires in a closed room were similar to those from pre-flashover fires of the same cushions in a room with the door open. The uncertainties in the post-flashover data are smaller due to the higher species concentrations and the more fully established upper layer from which the fire effluent was sampled. The uncertainty values are comparable to those estimated for the fractional effective dose calculations used to determine the time available for escape from a fire. The uncertainty in the yield data from the sofa, bookcase, and cable tests is sufficiently small to determine whether a bench-scale apparatus is producing results that are similar to or different from the real-scale results here. The use of Fourier transform infrared (FTIR) spectroscopy was shown to be a useful tool for obtaining concentration data of toxicants. However, its operation and interpretation is far from routine. The losses of CO, HCN, and HCl as they flowed down the corridor were found to be dependent on the combustible. The downstream to upstream concentration ratios varied from unity for some fuels to a factor of five smaller for others. The CO yield from two of the combustibles was significantly lower than the expected value of 0.2, which should be used in hazard and risk analyses. The accuracy of the results is verified, and a hypothesis is offered for the lower CO yield values.

Keywords: fire, fire research, smoke, room fire tests, fire toxicity, smoke toxicity

Disclaimer

Certain commercial entities, equipment, or materials may be identified in this document in order to describe an experimental procedure or concept adequately. Such identification is not intended to imply recommendation or endorsement by the National Institute of Standards and Technology, nor is it intended to imply that the entities, materials, or equipment are necessarily the best available for the purpose.

TABLE OF CONTENTS

vii

LIST OF FIGURES

LIST OF TABLES

EXECUTIVE SUMMARY

This report presents the methodology and results of Phase IIa of the International Study of Sublethal Effects of Fire Smoke on Survivability and Health (SEFS) project of the Fire Protection Research Foundation and the National Institute of Standards and Technology. The SEFS is a private/public fire research initiative to provide the scientific information for public policy makers to determine whether, when and how to incorporate the sublethal effects of heat and smoke in their fire safety decisions. The objective of this portion of the SEFS project is to establish a technically sound basis for assessing the accuracy of the bench-scale device(s) that will be generating smoke yield data for fire hazard and risk evaluation.

Estimation of the time people will have to escape or find a place of refuge in the event of a fire is a principal component in the fire hazard or risk assessment of an occupancy. Accurate assessment enables public officials and facility owners to provide a selected or mandated degree of fire safety with flexibility of design and confidence in the outcome, while imprecise assessment can result in increased cost and elimination of otherwise desirable building and furnishing products.

The computation in such an assessment involves the building design, the capabilities of the occupants, the potential growth rate of a fire and the spread rate of the heat and smoke, and the impact of the fire effluent (toxic gases, aerosols, and heat) on people in the fire vicinity. Increasing attention is being paid to the effects of the effluent on responders to a fire. The equations in ISO/TS 13571 now enable estimating the time available for escape. The approach to input data for such calculations is being addressed with the current work.

These data typically come from one of several, very different bench-scale combustors. Thus there can be diverse and perhaps conflicting data on fire effluent component yields available for any given product. Only one device (NFPA 269/ASTM E1678) has been validated against real-scale fire test data, and then only for post-flashover yields of the principal toxicants. This situation does not support either assured fire safety or marketplace stability. Thus, the need for a standard methodology for establishing the accuracy of these methods is critical to the credibility of fire hazard and risk assessments.

We report the results of room-scale fire tests. Various complex products were subjected to the key stages of a fire: well-ventilated flaming combustion and ventilation-limited (post-flashover) flaming combustion. Each test and fire phase included characterization of the fire and calculation of the yields of toxic gases and smoke. Since some of the toxicants might be removed from the inhalable environment, we estimated a degree of loss of each. Where the concentration was too low to be measured, an upper limit was estimated.

A future effort is being planned in which the same fuels are combusted in typical bench-scale apparatuses under combustion conditions appropriate to well-ventilated and ventilation-limited burning. The information generated in these real-scale fire tests then comprises the basis for assessing the accuracy of the yields from the various bench-scale devices.

Four combustibles were burned in a 2.44 m x 2.44 m x 3.66 m room whose only vent was an 0.76 m wide doorway leading to a 9.75 m long, open-ended corridor. In some tests, a panel was removed from the corridor ceiling 1.22 m downstream from the burn room, and the room effluent exhausted into a collection hood for heat release rate measurement. The optimal vent location was identified using Fire Dynamics Simulator (FDS) version 2.0, a computational fluid dynamics model employing large-eddy simulation techniques. The choice was for the nearest location at which the flame reaction was over before the effluent reached the vent location and where entrainment of dilution of the combustion products with corridor was minimized. Two tests were conducted with the door of the burn room closed.

The ignition modes and test configurations were selected to provide burning durations (under both pre-flashover and post-flashover conditions) that were long enough for substantive combustion product analyses. In some cases, adherence to realism was sacrificed to achieve this. The four combustibles were:

- "Sofas" made of up to 14 upholstered cushions supported by a steel frame. A 46 cm x 46 cm x 15 cm cushion consisted simply of a zippered cotton-polyester fabric over a block of foam. The elemental content of the cushions was (% by mass) 54.5 % C, 8.0 % H, 10.0 % N, 0.68 % Cl, 0.15 % P, and 26.7 % O. The mass of a cushion was about 1.1 kg and the heat of combustion was 24.4 MJ/kg ± 2.7 %. The California TB133 propane ignition burner faced downward, centered over the center of the sofa, about 10 cm above the top surface of the cushions. In all but two of the tests, the sofa was centered along the rear wall of the burn room facing the doorway. In two tests, the sofa was placed in the middle of the room facing away from the doorway to compare the burning behavior under different air flow conditions. Two of the sofa tests were in a closed room to examine the effect of vitiation. In these, an electric "match" was used to initiate the fires in the closed compartment tests.

- Particleboard (wood with urea formaldehyde binder) bookcases with a laminated vinyl finish. The bookcases were 1.83 m high x 0.91 m wide x 0.30 m deep. The back of the bookcase was a sheet of vinyl-laminated pressboard. The bookcase mass was *ca.* 27.5 kg. A diagonal length of steel angle iron was attached to the rear of the bookcases to prevent buckling and falling off the load cell during the test. The chemical analyses of the bookcases indicated a composition of 48.1 % C, 6.2 % H, 2.9 % N, 0.3 % Cl, and 42.6 % O. The heat of combustion was 18.16 MJ/kg ± 0.4 %. Early experiments with two bookcases side by side and the burner in between failed to sustain burning. As a result, two bookcases were placed in a "V" formation, with the TB133 burner facing upward approximately 30 cm under the lower shelves and 30 cm from the back of the "V."

- Rigid polyvinyl chloride (PVC) product sheet (a window frame material). Each test involved a single horizontal sheet of unplasticized PVC that was 0.71 m x 1.83 m x 7.9 mm in the room with burning bookcases. The elemental composition of the combustible portion of the sheet was 42.3 % C, 5.53 % H, and 52.2 % Cl. The measured heat of combustion was 16.17 kJ/kg ± 1.0 %.

- Household wiring cable, consisting of two 14 gauge copper conductors insulated with nylon and PVC, an uninsulated ground conductor, two paper filler strips, and an outer jacket of plasticized PVC. We estimated the fuel composition to be 45.8 % C, 6.2 % H, 1.62 % N, 25.2 % Cl, and 20 % O. The heat of combustion for the combustible fraction of the cable was 21.60 MJ/kg $\pm$ 0.6 %. Two 1.83 m long cable racks containing 3 trays each were constructed, with 30 kg of cable in each of the bottom two trays and 17 kg in each of the middle and top trays. The cable trays were placed parallel to the rear of the burn room. Twin 152 mm square propane ignition burners were centered under the bottom tray of each rack.

Supplies of each of the test fuels were stored for future use in bench-scale test method assessment.

The mass of each test specimen was monitored continuously. The concentrations of CO_2, CO and O_2 were monitored in the burn room and at three locations in the corridor using species-specific analyzers. Fourier transform infrared (FTIR) spectroscopy was used to monitor CO_2, CO, HCN, HCl, HF, HBr, NO, NO_2, H_2CO (formaldehyde), and C_3H_4O (CH_2=CH-CH=O, acrolein) at the upstream and downstream ends of the corridor. We were unsuccessful at determination of HCl, HCN, NO and NO_2 yields using a wet chemical technique. Soot was measured gravimetrically at the same two locations. All measurements were intended to be in the upper smoke layer, 30 cm from the ceiling. In the two tests with the doorway blocked, the effluent was sampled from the upper layer of the burn room. Additional measurements were made of the vertical temperature and pressure profiles in the doorway (for effluent flow calculation), CO, CO_2, and O_2 and flow in the exhaust hood (for heat release rate calculations), heat flux to the burn room floor (as a measure of flashover). All tests were videotaped within the burn room and down the corridor.

Following preliminary experiments to identify the ignition protocol, determine the mass of fuel needed to produce flashover, and measure the rate of heat release and rate of mass loss, 22 tests were then performed as follows:

- Three tests with an 8- or 12-cushion sofa. These tests did not proceed to flashover, but generated additional data for pre-flashover conditions.
- Five replicate tests with a 14-cushion sofa located against the back wall of the burn room, facing the open doorway. The intent was to provide an estimate of test repeatability.
- Two tests with the sofa against the back wall, but with the doorway blocked, to determine the effect of room vitiation.
- Seven tests of two bookcases each.
- Three similar bookcase tests with the rigid PVC sheeting product.
- Two tests of electric cable in the tray assembly.

The data from all the sensors (except the FTIR spectrometers) were collected electronically at 200 scans/s and smoothed to a rate of 1 sample/s. Channel markers kept track of the key events during a fire test. The FTIR data were recorded unsmoothed on a separate computer. All of the raw data (*ca.* 130 instruments, thousands of readings per instrument) from the tests reported here are to be available in a companion report. This will be in the form of spreadsheets and graphs.

For the open-door tests, the yields of the gases were determined by defining the pre- and post-flashover time intervals, determining the test specimen mass loss and the average volume fractions of the gases during those intervals, calculating the pre- and post-flashover yields of CO_2 from the above plus the calculated total mass flow through the doorway, and determining the yields of the other gases using their mass fraction ratios to the mass fraction of CO_2.

For the closed-room tests, we assumed that the upper layer was well mixed. The measured volume fractions of the gases and the ideal gas law were used to calculate the mass of each species in the upper layer. These were normalized to specimen mass loss, as a function of time.

For the PVC sheets, only post-flashover results were possible since the mass loss was negligible before flashover. It was assumed that all the HCl was from the PVC sheet and all the HCN came from the bookcases. Since the scatter in the CO and CO_2 yields was comparable to any differences between tests with and without the PVC sheet, yield data for these two gases from the PVC sheet were not calculable.

The uncertainty in the yield values results from the sensitivity of the yield to the selected time pre- or post-flashover time interval, the uncertainty in the specimen mass loss, the uncertainty in the species mass flow out the doorway (for open door tests), and the quality of the assumptions inherent in the calculation of the mass of product in the upper layer (for closed room tests). For the closed room tests, the uncertainty was further estimated by comparing the yield values from the early combustion with those from the pre-flashover segments of the open door sofa tests. The analysis of similar tests also structured the determination of uncertainty and repeatability.

Some of the data were not used because an instrument malfunctioned, the upper layer (containing the combustion products) did not fully envelop the sampling probe tips, or the concentration values were too close to the background levels.

We were able to obtain usable information using FTIR spectroscopic analysis. We note that its application to fire testing requires the constant attention of an experienced professional at a level well beyond the demands of the more traditional fire test instrumentation.

Initial checks on the consistency of the upstream post-flashover and late pre-flashover measurements showed the non-dispersive infrared (NDIR) and FTIR instruments gave similar concentrations of CO and CO_2 and low variability. Distinctly higher variability was found during the general pre-flashover burning periods for all tests. The pre-flashover sampling time period was adjusted such that the probe tip was sampling from the upper layer. The FTIR pre-flashover measurements were consistently smaller than those using NDIR for reasons not yet understood. The downstream pre-flashover CO measurements approached the detection limits of the analyzers. For the closed room tests, the early NDIR yields for CO_2 and CO were close to those for the open door sofa tests. As the fire progressed, the CO_2 yield decreased and the CO yield increased, as expected from burning in an increasingly vitiated atmosphere.

Many HCl and HCN measurements were very close to the background. Nonetheless, the data were sufficient to obtain reasonable post-flashover yield values and pre-flashover yield estimates for all three principal combustibles. The HCl concentration data for the PVC sheet were high enough to obtain a post-flashover HCl yield. The post-flashover HCl and HCN concentrations were also high enough to obtain estimates of the degree of loss of the compounds down the length of the corridor. The pre-flashover values had too high a degree of uncertainty for this use.

The equations in ISO/TS 13571 include additional gases to be included in estimating the time available for escape or refuge from a fire. The composition of the combustibles precluded the formation of some of these. Three key sensory irritants (NO_2, acrolein and formaldehyde) were not detected, thus establishing the upper limits of their presence at 100, 10, and 50×10^{-6} volume fraction, respectively. Analysis of these levels in light of their incapacitation concentrations from ISO/TS 13571 showed they would have had secondary contributions to incapacitation relative to the concentration of HCl, except in the case of the bookcases, which produced little HCl. This unimportance of secondary toxicants is consistent with the results of the animal experiments used to establish the N-gas hypothesis that attributes fire effluent lethality to a mall number of gases.

The following table presents the results of the measurements and calculations for yields of principal toxicants for both pre-flashover and post-flashover fires:

Gas	Fire Stage	Sofa	Bookcase	PVC Sheet	Cable
CO$_2$	Pre-fl.	1.59 ± 25 %	0.50 ± 50 %	--	0.120 ± 45 %
	Post-fl.	1.13 ± 25 %	1.89 ± 75 %	--	1.38 ± 15 %
CO	Pre-fl.	$1.44 \times 10^{-2} \pm 35$ %	$2.4 \times 10^{-2} \pm 55$ %	--	$5.5 \times 10^{-3} \pm 50$ %
	Post-fl.*	$5.1 \times 10^{-2} \pm 20$ %	$4.6 \times 10^{-2} \pm 30$ %	--	$1.48 \times 10^{-1} \pm 15$ %
HCN	Pre-fl.	$3.5 \times 10^{-3} \pm 50$ %	$4.6 \times 10^{-4} \pm 10$ %	--	$6.3 \times 10^{-4} \pm 50$%
	Post-fl.	$1.5 \times 10^{-2} \pm 25$ %	$2.5 \times 10^{-3} \pm 45$ %	--	$4.0 \times 10^{-3} \pm 30$ %
HCl	Pre-fl.	$1.8 \times 10^{-2} \pm 45$ %	$2.2 \times 10^{-3} \pm 75$ %	--	$6.6 \times 10^{-3} \pm 35$ %
	Post-fl.	$6.0 \times 10^{-3} \pm 35$ %	$2.2 \times 10^{-3} \pm 65$ %	$2.3 \times 10^{-2} \pm 85$ %	$2.1 \times 10^{-1} \pm 15$ %
NO$_2$	Pre-fl.	$< 7 \times 10^{-2}$	$< 2 \times 10^{-2}$	--	$< 4 \times 10^{-3}$
	Post-fl.	$< 1 \times 10^{-3}$	$< 1 \times 10^{-3}$	--	$< 1 \times 10^{-3}$
Acrolein	Pre-fl.	$< 8 \times 10^{-3}$	$< 2 \times 10^{-3}$	--	$< 4 \times 10^{-4}$
	Post-fl.	$< 1 \times 10^{-4}$	$< 1 \times 10^{-4}$	--	$< 1 \times 10^{-4}$
Formaldehyde	Pre-fl.	$< 2 \times 10^{-2}$	$< 2 \times 10^{-3}$	--	$< 8 \times 10^{-4}$
	Post-fl.	$< 8 \times 10^{-4}$	$< 4 \times 10^{-4}$	--	$< 7 \times 10^{-4}$

* See following discussion.

One check on the accuracy of the measurements was to compare calculated yields with the notional or maximum possible yields. Near-quantitative conversion of C and Cl in the fuel to CO_2 and HCl was expected. The post-flashover values of CO_2 from all three combustibles did just that, given the conversion of up to *ca.* 20 % of the carbon to carbonaceous smoke and CO. Under pre-flashover conditions, the yields were more variable. In the closed room tests, the yield began at about the notional level, then declined to about half that as room vitiation affected the completeness of combustion. The HCl yields were close to notional under post-flashover conditions for all the combustibles. Very low pre-flashover values for the electrical cable well reflect the known HCl reaction with the calcium carbonate filler in the cable jacket. While little of the nitrogen in the combustibles generally ended up in HCN, there was an over 10 % conversion from the post-flashover burning of the bookcases and cable.

The repeatability of the sofa tests was excellent: qualitative agreement of the shapes of the mass burning rate curves, similar global equivalence ratios, and low variability ($\pm$ 25 %) in the post-flashover yields of CO_2, CO, and HCN were within $\pm$ 25 % and are within $\pm$ 35 % for HCl. The pre-flashover yield values were repeatable to within a factor of two. For the sofa tests that did not reach flashover, the mass burning rate curves were also similar and the later pre-flashover CO_2, CO and HCl yields were repeatable to within $\pm$ 36 %, with the HCN yield repeatable to within $\pm$ 45 %. The yields from the two closed room sofa tests were repeatable to within $\pm$ 20 %.

The four cable tests showed qualitatively similar results. Post-flashover yield repeatability was typically $\pm$ 15 % to 30 %, with the pre-flashover repeatability somewhat higher but within a factor of two. For the four bookcase tests in which NDIR data were obtained, the post-flashover and pre-flashover yield repeatability values for CO_2 were *ca.* $\pm$ 75 % and $\pm$ 30 %, respectively; the CO values are *ca.* $\pm$ 30 % and $\pm$ 55 %. For the two bookcase tests for which we obtained FTIR data, the HCN post-flashover and pre-flashover yield repeatability values were *ca.* $\pm$ 45 % and 10 %, respectively, and the HCl values are 65 % and 75 %. The post-flashover HCl yields from the three PVC sheet tests spanned over an order of magnitude.

Of particular interest are the post-flashover yields of CO. A number of room-scale fire studies have indicated that the yield of CO is approximately 0.2 (g CO/g fuel consumed) and that this value is not very dependent on the combustible. In this study, the post-flashover CO yields from the cable fires approach this, with a mean of *ca.* 0.15 g/g. The sofas and bookcases generate about one quarter of the expected value.

We performed a number of checks to assure the accuracy of the CO yields. We verified the tests truly reached flashover. By comparison with CO levels within the burn room, we ascertained that the CO was not being oxidized in the secondary burning at the doorway. Experimental errors of a sufficient magnitude are highly unlikely, since two different types of analyzers with independent sampling lines produced comparable CO yields. The same calculations produced CO_2 yields near the notional limits, so there cannot be a missing factor in the data reduction.

A likely hypothesis is that large quantities of pyrolyzate are generated during flashover. These consume the limited available oxygen, forming CO, but leaving much of the organic matter unoxidized. As these gases reach the doorway and begin to entrain fresh air, more of the organic matter is oxidized to CO. Some of the CO is also oxidized to CO_2. Combined, these processes set up a dynamic situation where the observed $[CO]/[CO_2]$ ratio and the yield of CO depend on the degree of air-effluent mixing and the rate of cooling of the total flow. Since different fires and different stages of those fires are likely to be accompanied by differing degrees of CO formation and burnout, we suggest that for fire hazard and risk assessments, one should use the CO yield value of 0.2 g CO per g fuel consumed. Since bench-scale combustors typically used for generating toxic potency data generally do not have the potential for the secondary combustion processes described above, the 0.2 g/g value should also be used for assessing the accuracy of the data from such apparatus.

In summary, the repeatability of the yields values obtained in this study for three of the combustibles is sufficient for determination of whether a bench-scale apparatus is producing results that are similar to or different from the real-scale results here. The PVC sheet, from which only HCl yield data could be obtained, can only provide an indicator of appropriateness and then only for post-flashover simulation.

Since a large fraction of fire deaths result from post-flashover fires and since CO is always a major (if not the dominant) incapacitating toxicant, the repeatability results indicate an uncertainty in the fractional effective dose (FED) calculations that is comparable to the uncertainty in the equations themselves. The repeatability values should also be sufficient to determine whether a bench-scale apparatus is producing results that are similar to, or different from the real-scale results obtained in this study.

The loss of combustion products as they traveled down the corridor was quantified by the ratio of their upstream to downstream concentration ratios. Since CO_2 is inert, its ratio was used as a measure of dilution of the upper layer effluent with entrained lower layer air. Only post-flashover data were used due to the low values of the pre-flashover concentrations downstream. The losses of CO, HCN, and HCl were found to be dependent on the combustible. The downstream to upstream concentration ratios varied from unity for some fuels to a factor of five smaller for others. The cause of this is not understood. However, soot particles and aqueous aerosols are characterized by their number density, surface area, and hydrophilia. In these tests, only the soot mass was measured. It may well be that the smoke from the sofa and cable materials has a greater affinity for acid gases and CO than does the smoke from the bookcases and PVC sheet. However, for some other combustibles, loss factors of two to five beyond dilution are possible. Care should be taken not to extend these limited findings to other commercial products. Pending a comprehensive study of the relationship between smoke character and gas absorption, safety engineers are most likely to continue to assume there is no loss of toxicants, the more conservative approach.

The research was co-sponsored by the Alliance for the Polyurethane Industry, the American Plastics Council, DuPont, Lamson & Sessions, Underwriters Laboratories, and the Vinyl Institute under the aegis of the Fire Protection Research Foundation.

I. INTRODUCTION

This report presents the methodology and results of Phase IIa of the International Study of Sublethal Effects of Fire Smoke on Survivability and Health (SEFS) project of the Fire Protection Research Foundation and the National Institute of Standards and Technology. The SEFS is a private/public fire research initiative to provide the scientific information for public policy makers to determine whether, when and how to incorporate the sublethal effects of heat and smoke in their fire safety decisions. The objective of this portion of the SEFS project is to establish a technically sound basis for assessing the accuracy of the bench-scale device(s) that will be generating smoke yield data for fire hazard and risk evaluation.

Estimation of the time people will have to escape or find a place of refuge in the event of a fire is a principal component in the fire hazard or risk assessment of a facility. An accurate assessment enables public officials and facility owners to provide a selected or mandated degree of fire safety with confidence. An imprecise assessment can result in the regulator and/or designer applying large safety factors. These increase cost and can eliminate the consideration of otherwise desirable building and furnishing products.

Fire safety assessments now rely on some type of computation that takes into account such factors as the building design, the capabilities of the occupants, the potential growth rate of a fire and the spread rate of the heat and smoke, and the impact of the fire effluent (toxic gases, aerosols, and heat) on people in the fire vicinity.[1] Increasing attention is being paid to the effects of the effluent on responders to a fire.

The methodology for inclusion of fire effluent effects is presently *ad hoc* in nature, varying with the instance at hand and the person performing that portion of the safety assessment. The absence of a standard approach encourages conservatism while leaving questionable (both during the design process and in litigation following any mishap) the degree of safety provided.

It would thus bring an improved order to the construction and furnishing marketplace if there were a standard means of estimating the threat posed by fire effluent. This requires a calculation method and input data to support the calculations.

The first of these components is proceeding well:

- CFAST and other computer models of the movement of fire effluent throughout a facility have been in use for nearly two decades.[2] A number of laboratory programs and reconstructions of actual fires have given credence to the predictions.[3] These models calculate the temperature and combustion product concentrations as the fire develops. They can include equations for estimating when a person would die or is incapacitated, *i.e.*, is no longer available to effect his/her own escape.

- Devices such as the Cone Calorimeter[4] and related larger scale apparatus[5] are routinely used to generate information on the rate of heat release as a commercial product burns.

- With the adoption of ISO Technical Specification 13571, "Life Threat from Fires - Guidance on the Estimation of Time Available for Escape Using Fire Data," there now exist consensus equations for estimating the incapacitating exposures to narcotic gases,

1

irritant gases, heat and smoke.[6] Some of the basis for these equations lies in the prior effort under this project.[7]

The second of these components is addressed here. The equations in ISO/TS 13571 require data on the yields of the key combustion products from the various commercial products that might be involved in a fire. There has been, however, no standard methodology for routinely obtaining such yield data. Reliance on real-scale testing of commercial products is impractical for its expense per test and the vast number of commercial products used in buildings.

Rather, there are numerous bench-scale devices that are intended for generating chemical or physical measurements of smoke components. The combustion conditions and test specimen configuration in the devices vary widely, and some devices have wide flexibility in setting those conditions. Only one of these devices, used in both NFPA 269[8] and ASTM E1678[9], has been validated against real-scale fire test data, and then only for post-flashover yields of the principal toxicants. Meanwhile, ISO and IEC are proceeding toward standardization of a tube furnace, and ISO TC92 SC1 will be upgrading the analytical capability for the closed box test used by IMO and perhaps other similar devices. Thus, before too long there will be diverse and perhaps conflicting data on fire effluent component yields available for any given product. This situation does not support either assured fire safety or marketplace stability.

Thus, the need for a standard methodology for establishing the accuracy of these methods is critical to the accuracy and credibility of fire hazard and risk assessments.

The approach to be taken is to conduct a series of room-scale tests. Various complex products (as contrasted with single, homogeneous materials) are subjected to the key stages of a fire: well-ventilated flaming combustion and ventilation-limited (post-flashover) flaming combustion. The products are selected to generate the dominant toxicants. Each test and fire phase includes measurement of the mass yields (per mass of fuel consumed) of the principal combustion products contributing to lethal and sublethal effects of fire: heat, toxic gases, and particulates.

The list of toxicants to be monitored was: CO_2, CO, HCN, HCl, HF, HBr, NO, NO_2, H_2CO (formaldehyde) and C_3H_4O ($CH_2=CH-CH=O$, acrolein). Where the concentration was too low to be measured, an upper limit was estimated.

It is possible that some gases could be deposited on walls or soot as they travel away from the fire. Assuming no such losses could unduly penalize products containing, *e.g.*, halogenated additives in hazard analyses. Comparison of the concentrations at the upstream and downstream ends of the corridor provided some qualitative indication of the degree of loss of each gas.

A future effort is being planned in which the same fuels are combusted in typical bench-scale apparatuses under combustion conditions appropriate to well-ventilated and ventilation-limited burning. The information generated in these real-scale fire tests then comprises the basis for assessing the accuracy of the yields from the various bench-scale devices.

This report documents the room-scale experiments and the combustibles examined. It presents the combustion product yield data, their degree of repeatability, and the import of the findings.

II. EXPERIMENTAL INFORMATION

A. General Description

Four combustibles were burned in a room whose only significant vent was a doorway leading to a corridor; the downstream end of the corridor was unconfined. There were two types of tests:

- The first type was used to scope the burning behavior of the fuel and to guide the protocol for the second type of tests. A large hole in the corridor ceiling enabled measurements of CO_2, CO and O_2 concentrations to be made in the exhaust stack.

- The second type of test was used to determine the yields of the toxic gases and to determine the extent to which the more reactive ones were lost to soot or wall surfaces. The concentrations of the above gases were measured at three locations in the upper layer of the corridor and in the upper layer of the burn room. The concentrations of other gases of toxicological interest were measured at two locations in the corridor using Fourier transform infrared (FTIR) spectroscopy. In order to support the information on acid gases obtained using FTIR spectroscopy, we attempted (unsuccessfully) independent determination of HCl, HCN, NO and NO_2 yields using a wet chemical technique.

Yields of the toxic gases were calculated using the consumed mass of the fuel, gas concentrations measured in the corridor, and data regarding the flow down the corridor. The loss of product gases was estimated from the difference between the downstream to upstream concentration ratio of the gas and the same ratio for CO_2, whose concentration was presumed to change only by dilution. Gross measurements of soot density were made in order to enable future analysis of the observed losses.

B. Fire Test Configuration

1. Room Construction

The tests were conducted in the two-compartment assembly shown schematically in Figure 1 and photographically in Figures 2 and 3. The interior of the burn room was 2.44 m wide, 2.44 m high, and 3.66 m long (8 ft x 8 ft x 12 ft). The attached corridor was 9.75 m long (32 ft) and of width and height similar to the burn room. A doorway 0.76 m (30 in) wide and 2.0 m (80 in) high was centered in the common wall. The downstream end of the corridor was fully open, *i.e.*, there was no end wall. The entire assembly was elevated 76 cm (30 in) on cinder block supports.

During the first scoping tests, the walls and ceiling of the burn room and corridor were constructed of two layers of 1.27 cm (0.5 in) thick gypsum wallboard over wooden studs. After the tenth heat release test and prior to the first performance test (BW1, see Section II.D), the surface layer of gypsum board covering the walls and ceiling of the burn room was replaced with a single layer of calcium silicate board of the same thickness. As the test series progressed, this layer was spackled or replaced to keep smoke and heat leakage to a minimum.

Figure 1. Schematic of the Room-corridor Test Fixture

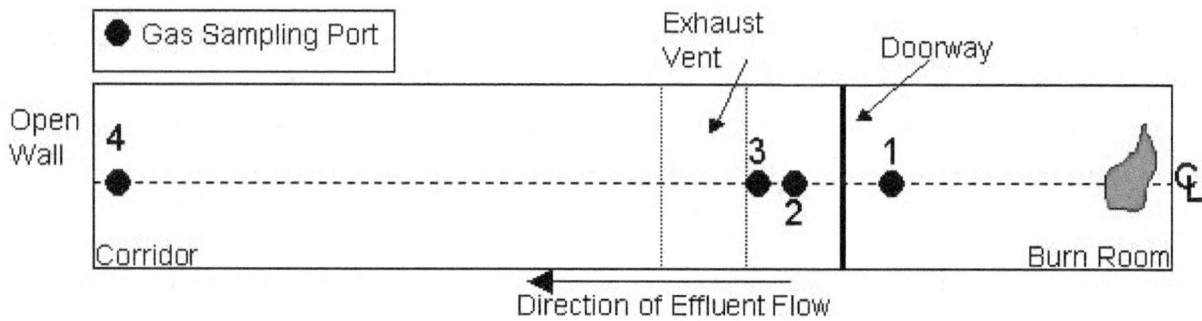

Figure 2. Photograph of the Exterior of the Room-corridor Test Fixture

Figure 3. Photograph of the Interior of the Room-corridor Test Fixture

For the tests designed for the measurement of rate of heat release (containing a "Q" in the test designation, Section II.D), a 1.22 m wide and 2.44 m long (4 ft x 8 ft) corridor ceiling panel was removed from the corridor 1.22 m (4 ft) downstream from the burn room wall (Figure 1). The room effluent exhausted through this vent into a large (6 m x 6 m aperture) collection hood, which was fit with the instrumentation for heat release rate measurement.[10] During the production tests, this vent was sealed, and the room effluent flowed the full length of the corridor to a large, uninstrumented exhaust hood.

The optimal location for the vent was identified using Fire Dynamics Simulator (FDS) version 2.0, a computational fluid dynamics model employing large-eddy simulation techniques.[11] Four locations in the corridor ceiling were investigated. In each case, the vent width was the full width of the corridor, 2.44 m (8 ft), and the length was 1.22 m (4 ft). The fire in the calculations was a sofa fire that produced flashover. The nature and behavior of the fire plume exiting the doorway was calculated for each vent location, with a fifth computation simulating the same fire with no vent opening.

- For the computation with no vent opening, the flames extending out the doorway impinged on the ceiling within about 1 m of the doorway plane and were quenched.

- For the exhaust vent location 0 m to 1.22 m from the doorway, flames extended through the doorway and into the vent opening toward the calorimetry measurement hood. This unquenched chemistry could result in chemical composition of the effluent (and thus a rate of heat release) different from the later tests with no vent opening, an undesirable outcome.

- For the exhaust vent location between 1.22 and 2.44 m downstream of the doorway, the modeling showed that the flame extension would not continue past 1.22 m downstream, thus only quenched effluent flowed through the vent.

- The model results for the exhaust vent location further downstream showed the same desired flame quenching phenomena. However, the exhaust gases would travel further than the previous case, resulting in increased entrainment and mixing. The increased entrainment and mixing results in an undesirable temporal averaging of the heat release rate measurement.

Thus, the exhaust vent for the corridor was located between 1.22 m and 2.44 m downstream of the doorway.

Figure 4 is a visualization of the FDS simulations of a sofa fire with the exhaust vent closed. The orange area (or dark gray, if viewed in black and white) of the plume represents the surface of the flame sheet where the mixture fraction is 1, *i.e.*, where fuel and oxygen are assumed to react stoichiometrically. Note that the flame reaction is over before the effluent reaches the 1.22 m to 2.44 m vent location, while entrainment of corridor air has yet to dilute the combustion products appreciably.

Two of the tests were carried out with the doorway blocked. In those cases, there was no venting of the effluent. Rather, the effluent accumulated in the upper layer of the room, from which it was sampled.

Figure 4. Simulation of a Sofa Fire Using FDS

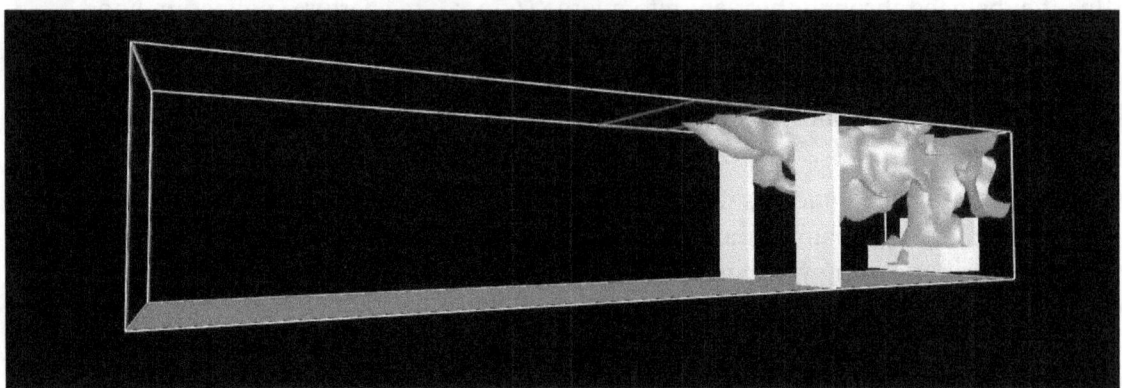

2. Load Cells

Two load cells (described in Section II.E.3 below) were used to measure the specimen mass loss during the tests. The load cells were placed on the floor of the test bay below the burn room. The combustible was placed on a large metal pan that was in turn supported on a frame that transmitted the mass through holes in the burn room floor to the load cell.

3. Sampling Ports

Gases and soot were sampled at some or all of four locations. The tips of the single probes and the middle of the four-probe arrays were located on the corridor/burn room centerline, approximately 30 cm (1 ft) from the ceiling with the intent to avoid sampling from within a stagnant boundary layer but still capture combustion products from early, low-momentum effluent flows. The tubing lengths of the two four-probe arrays were parallel, their tips forming a diamond 100 mm high and wide. The probes are shown in Figure 5 and described in Table 1. The axial locations were:

- Single probe 1 m (3.3 ft) inside the burn room door. The desire was to obtain information on the fixed gases (CO_2, CO, and O_2) in the well-mixed, upper layer. For the tests with the burn room door closed, room gas was extracted from a similar adjacent port for FTIR analysis.

- Four-probe array (see Section II.E.5 for more complete description), nominally 1 m (3.3 ft) outside the burn room doorway. This location was selected to be in the upstream end of the quenched doorway jet, *i.e.*, in a location where minimal entrainment of corridor air and dilution of the combustion products would have occurred following their leaving the burn room. For the more intense fire stages, the flames were not always quenched at this location.

- Single probe 2.1 m (6.6 ft) downstream from the burn room doorway (30 cm (1 ft) upstream of the heat release rate vent). The purpose of measurement at this location was to characterize the composition of the fixed gases just before they reached the exhaust vent.

- Four-probe array nominally 9.4 m (30.8 ft) from the burn room door or approximately 1 m (3.3 ft) upstream from the open end of the corridor. This location was selected in order to be as far down the corridor as possible, yet minimize edge effects at the end of the corridor.

Table 1. Description of Four-probe Sampling Arrays

Probe Designation	Probe Location	Function
T	Top	Pre-flashover soot
B	Bottom	Post-flashover soot
U	Upstream	FTIR analysis
D	Downstream	Fixed gases

Figure 5. Photographs of Sampling Probes

5a. Corridor Interior

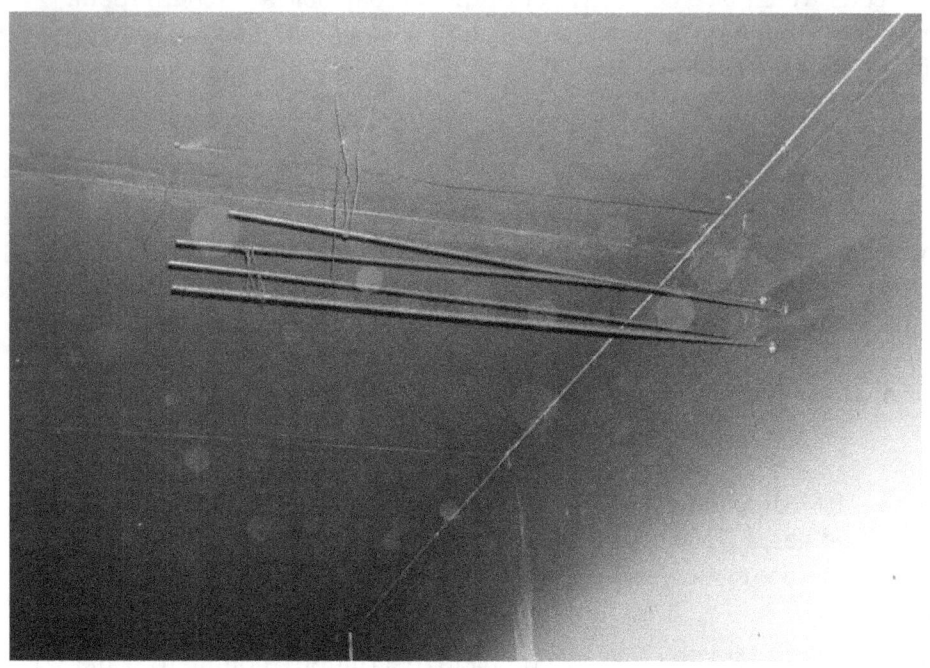

5a. Corridor Exterior

4. Ignition Burners

Two different propane burners were used as ignition sources for the test series. The first burner, the one for testing mattresses under California Technical Bulletin 133, was used for the sofa and bookcase tests. The burner is described in detail elsewhere.[12] Briefly, it consists of a perforated square "ring" with an outer dimension of 0.25 m (0.8 ft) attached to a supply tube at the center of one side of the square. The burner ring and supply tube were made of 12.7 mm (1/2 in) diameter stainless steel. The supply tube was connected to a compressed gas cylinder containing propane via a 12.7 mm (0.5 in) flexible supply line. A valve and flowmeter were located just downstream of the propane cylinder.

The electrical cable was ignited using two 152 mm (6 in) square sand-filled steel burners connected at the centers of their bottoms by a 12.7 mm (0.5 in) steel pipe. Propane was supplied to the burners from a compressed gas cylinder through a flowmeter installed in the 12.7 mm (0.5 in) supply line at a flow of 0.024 m^3/min (50 ft^3/hr).

To initiate the fires in the closed compartment tests, we fabricated an "electric match" as follows. The cover of a cardboard matchbook was bent backward, and a loop at one end of a length of small gauge nichrome wire was inserted through the sulfur ends of the matches. The other end of the wire was attached to a switched power source. When the switch was closed, the wire heated quickly, igniting the matchbook within 2 s. Since the matchbook burns out quickly, it was surrounded by a folded piece of paper to extend the flaming for approximately 20 s, ensuring ignition of the surrounding material.

C. Test Specimens

Four fuels were selected for diversity of physical form, combustion behavior, and the nature and yields of toxicants produced:

- "Sofas" made of upholstered cushions supported by a steel frame. The fire retardant in the cushion padding contains chlorine atoms. Thus this fuel would be a source of CO_2, CO, HCN, HCl, and partially combusted organics.
- Particleboard bookcases with a laminated vinyl finish. This fuel would be a source of CO_2, CO, partially combusted organics, HCN and HCl.
- Rigid PVC product sheet (window frame material). This fuel would be a source of CO_2, CO, HCl, and partially combusted organics.
- Electric power cable in a 3-D array of horizontal trays. This fuel would be a source of CO_2, CO, HCl, and partially combusted organics.

Photographs of these appear in Section II.D.

Specimens of the principal components of each fuel were sent to an independent testing laboratory to characterize their chemical nature. The data were obtained by combusting small (*ca.* 10 mg) samples and measuring the combustion products. Generally single analyses were performed on three samples taken from different pieces of the combustibles. Since there was extensive unburned residue from the cable fires and since there was a possibility that the residue chemistry might differ significantly from the composition of the unburned product, three samples of the char from a single fire were sent to the test laboratory. They performed measurements on duplicate specimens from each of the three samples. The analytical chemical data are shown in Table 2. The elemental composition of the component materials in the fuels is shown in Table 3. Additional data on the heats of combustion (triplicate samples) are shown in Table 4.

The details of the composition of the fuels and their test configurations are discussed below. The ignition modes and test configurations were selected to provide burning durations (under both well-ventilated and ventilation-limited conditions) that were long enough for accurate combustion product analyses. In some cases, adherence to realism was sacrificed to achieve this. Some of the sofas were burned in two different orientations to estimate the effect of fuel location on combustion product yields.

Supplies of each of the test fuels were stored for future use in bench-scale test method assessment.

Table 2. Elemental Analysis of Fuels

Sample	Mass %														
	C	H	N	Cl	Ca	Pb	Al	Sb	P	Sn	Ti	Total	Δ*	O**	Remainder
Particle Board, with laminate	46.89	6.70	2.68	0.26	n	n	n	n	n	n	n	56.53	43.47		
	46.56	6.68	3.35	0.24	n	n	n	n	n	n	n	56.83	43.17		
	47.12	6.60	2.76	0.26	n	n	n	n	n	n	n	56.74	43.26		
Mean value	46.86	6.66	2.93	0.25								56.70	43.30	42.6	0.7
Standard deviation	0.28	0.05	0.37	0.01								0.15	0.15		
Pressboard, with laminate	43.04	6.12	0.21	0.14	n	n	n	n	n	n	n	49.51	50.49		
	43.12	6.08	0.21	0.15	n	n	n	n	n	n	n	49.56	50.44		
	42.73	6.20	0.18	0.14	n	n	n	n	n	n	n	49.25	50.75		
Mean value	42.96	6.13	0.20	0.14								49.44	50.56		
Standard deviation	0.21	0.06	0.02	0.01								0.17	0.17		
Cushion fabric	47.23	6.23	0.18	n	n	n	n	n	n	n	n	53.64	46.36		
	48.12	6.10	0.19	n	n	n	n	n	n	n	n	54.41	45.59		
	47.38	5.99	0.20	n	n	n	n	n	n	n	n	53.57	46.43		
Mean value	47.58	6.11	0.19									53.87	46.13	46.5	0.4
Standard deviation	0.48	0.12	0.01									0.47	0.47		
Cushion padding	56.38	8.48	12.58	0.95	n	n	n	n	0.20	n	n	78.59	21.41		
	56.33	8.58	12.50	0.90	n	n	n	n	0.15	n	n	78.46	21.54		
	56.36	8.53	12.46	0.71	n	n	n	n	0.21	n	n	78.27	21.73		
Mean value	56.36	8.53	12.51	0.85					0.19			78.44	21.56	25	-3.5
Standard deviation	0.03	0.05	0.06	0.13					0.03			0.16	0.16		
PVC sheet	35.98	4.64	<0.10	43.23	3.01	n	n	n	n	0.38	5.92	93.16	6.84		
	35.87	4.57	<0.10	43.31	3.01					0.36	5.87	92.99	7.01		
	36.00	4.57	<0.10	43.34	3.03					0.37	5.84	93.15	6.85		
Mean value	35.95	4.59		43.29	3.02					0.37	5.88	93.10	6.90	7.6	-0.7
Standard deviation	0.07	0.04		0.06	0.01					0.01	0.04	0.10	0.10		

11

Sample	Mass %											Total	Δ *	O**	Remainder
	C	H	N	Cl	Ca	Pb	Al	Sb	P	Sn	Ti				
Cable jacket	40.83	5.07	<0.10	26.77	10.42	< 0.05						72.67	27.33		
	40.94	5.20	<0.10	26.53	10.18	< 0.05						72.67	27.33		
	40.87	5.15	<0.10	26.68	10.24	< 0.05						72.70	27.30		
Mean value	40.88	5.14		26.66	10.28	< 0.05						72.68	27.32	16.7	10.6
Standard deviation	0.06	0.07		0.12	0.10							0.02	0.02		
Wire insulation	48.25	6.73	2.39	26.04			0.80	0.62				83.41	16.59		
	48.20	6.98	2.65	26.08			0.81	0.62				83.91	16.09		
	48.57	6.82	2.40	26.22			0.72	0.63				84.01	15.99		
Mean value	48.34	6.84	2.48	26.11			0.78	0.62				83.78	16.22	9.2	7.0
Standard deviation	0.20	0.13	0.15	0.09			0.04	0.00				0.32	0.32		
Cable filler	42.58	6.65	<0.10	n								49.23	50.77		
	42.42	6.84	<0.10	n								49.26	50.74		
	42.72	6.80	<0.10	n								49.52	50.48		
Mean value	42.57	6.76										49.34	50.66	38.4 (C) 49.0 (H)	
Standard deviation	0.15	0.10										0.16	0.16		
Cable residue	18.39	2.30	0.20	22.99								43.88	56.12		
				25.42											
	19.03	2.45	0.21	27.76								49.45	50.55		
				28.62											
	17.91	2.47	0.14	30.00								50.52	49.48		
				27.87											
Mean value	18.44	2.41	0.18	26.96								47.95	52.05		
Standard deviation	0.56	0.09	0.04	2.51								3.57	3.57		

*   [1 - Σ (mass %) of listed elements]

** See following text for estimation methods

12

Table 3. Elemental Analysis of Fuel Components

Sample	Mass %												
	C	H	N	Cl	Ca	Pb	Al	Sb	P	Sn	Ti	Ash	O
Wood	49.0	6.1	0.2									0.5	44
Paper	49.0	6.1	0.2									0.5	44
Urea formaldehyde	33.3	5.6	38.9										22.2
PVC	38.4	4.8		56.7									0
Dioctyl phthalate	73.8	9.8											16.4
Melamine	28.6	4.8	66.7										0
Cotton (= cellulose)	44.5	6.2											49.3
Polyethylene terephthalate	62.5	4.2											33.3
Nylon 6,6	64	9.3	12										14
Nylon 6	66	10.2	11										13
FPU	57.6	5.6	11.2										25.6

Table 4. Heats of Combustion of Fuels

Sample	ΔH_c (MJ/kg)			Mean	σ
Particle Board, with laminate	18.24	18.17	18.07	18.16	0.07
Pressboard, with laminate	16.48	16.18	16.26	16.31	0.03
Cushion fabric	18.17	17.96	17.94	18.02	0.10
Cushion padding	26.09	26.02	26.12	26.08	0.04
PVC sheet	16.67	16.48	1.27	16.47	0.17
Cable jacket	18.30	18.41	18.36	18.36	0.04
Wire insulation	23.39	23.33	23.45	23.39	0.06
Cable filler	17.01	17.00	17.00	17.00	0.00
Cable residue	Did not ignite				

1. "Sofas"

These were made of arrays of upholstered cushions supported by a steel frame. The cushions consisted of a zippered fabric over a block of foam, with no interliner or other components. The finished cushions were each nominally 46 cm x 46 cm x 15 cm (18" x 18" x 6").

The fabric was described by the supplier as a cotton-polyester blend with no added fire retardant. We assumed that the polyester was a terephthalate, the formulation typically used in fabrics. These two components contain only carbon, hydrogen and oxygen, so the source of the nitrogen in the sample analyses (Table 2) is unknown. From the carbon fraction of the polymers and the sample analysis data in Table 2, we estimate that the fabric is about 82 % cotton by weight.

The foam was described as a flexible polyurethane formulation containing melamine and a chlorinated phosphate ester fire retardant. Based on this information, we requested elemental analyses for C, H, N, P and Cl. Adding stoichiometric masses of oxygen from the phosphate and foam (assuming a TDI-polyol formulation) components, we were able to estimate the mass percentage of oxygen in the components. As can be seen from Table 1, this estimation accounts for the specimen mass to within *ca.* 3 %. The formulation of the foam is thus presumed to be well defined.

We separated five of the cushions into their fabric and foam components and weighed them. The masses of the components and the cushions are shown in Table 5. Since the cushions appeared to burn evenly (*i.e.*, the fabric was generally not burned away well before the foam was) and since they were virtually consumed in the tests (Section IV.A), we presumed that the elemental composition of the fuel was steady during the tests.

Table 5. Mass (g) and Mass Fraction of Cushion Components

Sample	Fabric	Padding	Sum
1	236 (0.202)	933 (0.798)	1169
2	237 (0.201)	944 (0.799)	1181
3	240 (0.206)	925 (0.794)	1165
4	239 (0.202)	942 (0.798)	1181
5	244 (0.212)	907 (0.788)	1151
Mean	239 (0.205)	930 (0.795)	1169
σ	3 (0.004)	12 (0.004)	

Given the fractions of the two components, we then estimated the cushion composition (mass fraction) to be:

$$C: \quad 0.545 \pm 1\%$$

$$H: \quad 0.080 \pm 1\%$$

$$N: \quad 0.100 \pm 1\%$$

$$Cl: \quad 0.0068 \pm 16\%$$

$$P: \quad 0.0015 \pm 17\%$$

$$O: \quad 0.267 \pm 4\%$$

The fuel mass of 8-, 12-, and 14-cushion sofas was approximately 9 kg, 14 kg, and 16 kg, respectively. Using the heat of combustion for the components (Table 3) and the above component fractions, the derived value for the heat of combustion for the cushions is 24.4 MJ/kg $\pm 3\%$.

The steel frame was fabricated of 4 cm (1.5 in) angle iron. A steel plate was placed on the seat to prevent collapse of the seat cushion. To stabilize the back cushions, a similar plate was placed on the back of the frame and the back was angled backwards at *ca.* 3 ° from the vertical. To prevent the back cushions from toppling (possibly off the weighing platform), they were attached to the frame with heavy unclad wire.

The ignition burner was placed facing downward, centered over the center of the sofa, about 10 cm above the top surface of the cushions. The propane flows are indicated in the description of the individual tests (Section IV.A).

In all but two of the tests, the sofa was centered along the rear wall of the burn room, approximately 7 cm from the wall, facing the doorway. In two of the preliminary tests, the sofas were placed in the middle of the room facing away from the doorway. The intent had been to compare the burning behavior under different air flow conditions. However, the resources were not available to perform the fully instrumented tests needed to complete this assessment.

The initial experiments involved a sofa consisting of a four-cushion seat and a four-cushion back. These did not result in flashover of the test room, except for the first such test, in which the paper lining of the wall covering ignited and provided the additional heat release needed for flashover. The results from this one test were not used further due to this mixing of fuels in an unknown ratio. Observations of these eight-cushion tests indicated that the center cushions were consumed before the end cushions were fully involved. Accordingly, one test (SW3) was performed with doubled seat cushions in the center and "armrest" cushions at either end. The space under the outer seat cushions was filled with drywall. This 12-cushion test marginally failed to produce flashover. The remaining tests were conducted with doubled seat cushions in the center, a second tier of back cushions, no armrests, and a modified test frame. The space under the outer seat cushions was filled with drywall. These 14-cushion arrays resulted in an acceptable pre-flashover burning period, flashover, and an acceptable post-flashover burning period before the fuel began to burn out.

2. Bookcases

The dimensions of the bookcases were 1.83 m high x 0.91 m wide x 0.30 m deep (6 ft x 3 ft x 1 ft). Each bookcase contained one fixed and one adjustable shelf (0.95 m and 0.72 m from the base of the bookcase, respectively). The finished board stock of the frame and the shelves was 25.4 mm (1") thick. The back of the bookcase was a sheet of vinyl-laminated pressboard approximately 5 mm in thickness. The typical mass of a bookcase was 27.5 kg. A diagonal length of steel angle iron was attached to the rear of the bookcases to prevent buckling and falling off the load cell during the test.

With the exception of the chlorine content of the particleboard, the elemental compositions of the components were similar. Since the mass fraction of the back panel was small and since it tended to burn away extensively before the combustion of the particleboard was established, we assumed that the test material was essentially the laminated particleboard.

Samples of the sawdust from cutting the shelves were collected and sent for analysis for C, H, N, and Cl. We assumed that there was no fire retardant additive and thus looked for no additional elements. We assumed that the nitrogen came mainly from the urea formaldehyde binder, with a small contribution typical of wood. Using the measured mass fraction of nitrogen in the bookcase sample, we estimated that the composite is about 7 % urea formaldehyde resin by mass. Since chlorine was present in the elemental analysis, we assumed that the laminated finish was polyvinylchloride. Using the mass fraction of chlorine in the bookcase sample, we estimate that the composite is about 0.2 % PVC by mass. We obtained an empirical composition of wood from the published literature. After removing the mass fraction of the (non-combustible) ash, this led to an estimate of the fuel composition to be:

C:	$0.481 \pm 0.6 \%$
H:	$0.062 \pm 0.8 \%$
N:	$0.029 \pm 13 \%$
Cl:	$0.0030 \pm 4 \%$
O:	$0.426 \pm 1 \%$

Given the possible variation in the elemental composition of different sources of woods, this is in good agreement with the analytical results from the test laboratory (Table 2). We concluded that there were no significant additional components in the bookcases.

We again assumed that the atomic composition of the fuel was steady during the tests and that the char mass was a small fraction of the unburned fuel. We then used the elemental analysis results from Table 2 to compute the notional gas yields. The heat of combustion for the bookcase was equated to that for the particleboard (Table 4) as 18.16 MJ/kg $\pm$ 0.4 %.

Early experiments with two bookcases side by side and the burner in between showed that the burner could ignite the pressboard back but not the particleboard. As a result, two bookcases were placed in a "V" formation (*ca.* 10° angle) to provide radiative enhancement and to trap the heat from the incipient fire. The rear edges of the bookcases were almost touching each other and the front edges were approximately 30 cm apart. The rear of the "V" was about 5 cm from the rear wall of the burn room. The ignition burner was the same as that used for the cushions.

It was placed facing upward approximately 30 cm under the lower shelf and 30 cm from the back of the "V." These arrays resulted in a lengthy pre-flashover burning period, flashover, and an acceptable post-flashover burning period before the burning rate diminished.

3. **Rigid PVC sheeting**

This was described by the supplier to be of a composition similar to that used for vinyl window framing. Each test involved a single sheet of unplasticized PVC that was 0.71 m x 1.83 m x 7.9 mm (28" x 72 " x 0.31").

The manufacturer provided the following approximate composition guidance: 75 % PVC resin, 7.5 % $CaCO_3$, 2 % Sn stabilizer, 7.5 % TiO_2, 0.5 % process aid, 4 % acrylic impact modifier, 3.5 % pigments. We thus requested analyses for C, H, Cl, Ca, Sn, and Ti. Using the empirical formulas for the metal salts, we estimated the mass fraction of oxygen in the specimens. As shown in Table 2, this estimation accounts for the specimen composition within 1 %, and we concluded that there were no additional components of significant contribution.

We again assumed that the atomic composition of the fuel was steady during the tests and that the organic residue was a small fraction of the unburned fuel. Thus, for estimating the notional yields of product gases, we used the mean values of the elemental analyses as received from the testing laboratory, corrected for the non-volatile inorganic additives to obtain:

$$C: \quad 0.423 \pm 0.2 \%$$

$$H: \quad 0.0553 \pm 0.9 \%$$

$$Cl: \quad 0.522 \pm 0.2 \%$$

The measured heat of combustion for the PVC sheet was 16.17 kJ/kg $\pm$ 1.0 %.

The PVC sheeting was only combusted in concert with burning bookcases. The sheet was supported horizontally on an angle iron frame 0.81 m (32 in.) above the floor of the front of the burn room. The length of the sheet was parallel to the front of the room and 0.81 m (32 in) from it. Radiative ignition occurred as the combustion of the bookcases approached flashover.

4. **Electric power cable**

This was typical of the product used for household wiring. It had two 14 gauge copper conductors insulated with nylon and PVC, an uninsulated ground conductor, two paper filler strips, and an outer jacket of plasticized PVC.

We provided the testing laboratory with separated samples of the jacket, the wire insulation, and the filler material. Based on information from the manufacturer, we requested analyses for C, H, N, Cl (insulation and jacket only), Al and Sb (insulation only), and Ca and Pb (jacket only). We instructed the test laboratory to cut cylindrical slices of the insulation cylinders to promote even sampling of the nylon and PVC components.

To estimate the oxygen fraction of the cable jacket, we assumed that the plasticizer in the PVC was dioctyl phthalate (DOP), and that no additional fire retardants had been added. We estimated the PVC fraction (*ca.* 0.47) from the chlorine fraction in the sample (Table 2) relative to the Cl fraction in pure PVC (Table 3). Similarly, we estimated the $CaCO_3$ fraction (*ca.* 0.26) from the Ca fractions in the two tables. We then obtained the DOP fraction (*ca.* 0.27) by difference. From the chemical formulations of the three components and these composition fractions, we estimated the O mass %. As can be seen from the rightmost column of Table 2, there is clearly an unaccounted component in the jacket. Since the relative organic component fractions and the elemental composition are self-consistent, we expect that there is an additional inorganic filler present.

A similar analysis was performed for the wire insulation. The PVC fraction (*ca.* 0.47) was estimated as above, the nylon fraction (*ca.* 0.24) was estimated from the N content, the aluminum trihydrate content (*ca.* 0.037) from the Al content, and the antimony oxide fraction (*ca.* 0.0074) from the Sb content. The DOP fraction (*ca.* 0.19) was obtained by making the sum of the carbon contributions from the PVC, nylon, and DOP components equal the chemical analysis results. There is evidence of an unidentified component, most likely an inorganic filler.

We did a similar analysis for the paper. Since the analyzed nitrogen fraction was very small, we assumed that the sample consisted only of C, H, and O. We then calculated the O fraction using the paper chemistry from Table 3 and the elemental analysis from Table 2. Depending on whether we balanced the carbon content of the filler or the hydrogen content, we obtained two different results, as shown in Table 2. However, since the paper constituted a small fraction of the total mass, this was not pursued further.

We weighed the components of samples from five different reels of cable. Each sample was about 30 cm in length. The results are shown in Table 6.

Table 6. Mass (g) and Mass Fraction of Electrical Power Cable Components

Sample	Insulation	Wire	Paper	Jacket	Sum
1	14.3 (0.518)	5.9 (0.214)	0.9 (0.033)	6.5 (0.235)	27.6
2	14.5 (0.522)	5.9 (0.212)	0.9 (0.032)	6.5 (0.234)	27.8
3	14.7 (0.521)	6.0 (0.213)	0.9 (0.032)	6.6 (0.234)	28.2
4	13.8 (0.502)	5.8 (0.211)	0.9 (0.033)	7.0 (0.255)	27.5
5	13.7 (0.517)	5.6 (0.211)	0.9 (0.034)	6.3 (0.238)	26.5
Mean	14.2 (0.516)	5.8 (0.212)	0.9 (0.033)	6.6 (0.239)	27.5
σ	0.4 (0.007)	0.1 (0.001)	0.0 (0.0007)	0.2 (0.007)	
Combustible Fraction	0.655 ± 0.009		0.042 ± 0.009	0.303 ± 0.009	0.788

Table 2 shows that the chlorine content of the char was similar to that of the unburned fuel. We thus assumed that the elemental composition of the consumed fuel was steady during the tests. After removing the mass fraction of the (non-combustible) copper conductor and the inorganic fraction, this led to estimates of the fuel composition to be:

$$C: \quad 0.576 \pm 0.5\,\%$$
$$H: \quad 0.080 \pm 1.5\,\%$$
$$Cl: \quad 0.323 \pm 0.4\,\%$$
$$N: \quad 0.021 \pm 6\,\%$$

We used these results to compute the notional gas yields.

We assumed that the copper remained in its initial, non-oxidized state. Using the heats of combustion for the components (Table 4) and the above composition fractions, the derived value for the heat of combustion for the combustible fraction of the cable is 21.60 MJ/kg $\pm$ 0.6 %.

To determine the size of the cable array needed to bring the burn room to flashover and to sustain post-flashover burning for about 3 minutes, we used data from Dey [13] on the heat release rate per unit surface area of a cable tray and estimated values of the heat of combustion (25 MJ/kg) and specific mass loss rate (3 g/m^{2}s). The calculation indicated that six trays, each 1.83 m by 0.31 m in surface and containing 5 layers of cable would suffice. Thus, two cable racks containing 3 trays each were constructed.

The cable was cut to lengths of 1.83 m $\pm$ 0.03 m. The bottom two trays were disproportionately loaded since they would ignite first and should not burn out before all six trays were aflame. The bottom two trays held approximately 30 kg of cable each, while the middle and top trays held about 17 kg each. As expected, these arrays resulted in an ample pre-flashover burning period, flashover, and an acceptable post-flashover burning period before the burning rate diminished.

The cable trays were placed parallel to the rear of the burn room. The rear tray was 300 mm from the rear wall; the space between the trays was 15 cm. The burner centers were 380 mm (1.25 ft) apart and were centered under the bottom tray of each rack.

D. Test Plan

For each product type, a series of preliminary experiments (Table 7) was performed to:

- Identify ignition protocols and fuel distribution to produce the desired burn period of two to three minutes before the combustion would become ventilation limited.
- Determine or verify the mass of fuel needed to produce flashover in the room and sustain it for two to three minutes.
- Measure the rate of heat release and rate of mass loss, enabling calculation of the effective heat of combustion from these fuel packages.

Tests were also conducted to ascertain any differences in the rate of heat release and combustion efficiency between:

- A sofa in the rear of the burn room facing the doorway and
- A sofa in the center of the room and facing away from the doorway.

For these 11 tests, as noted in Section II.A, a panel was removed from the corridor ceiling for determination of the rate of heat release (Section III.A.)

Twenty-two room-scale fire tests were performed with the ceiling hole closed (Table 7). These tests are categorized as follows:

- SW1-3: Three tests with an 8- or 12-cushion sofa located against the back wall of the burn room, facing the open doorway. These tests did not proceed to flashover, but generated data for pre-flashover conditions. (Figures 6-7)

- SW10-14: Five replicate tests with a 14-cushion sofa located against the back wall of the burn room, facing the open doorway. The intent was to provide an estimate of test repeatability. (Figure 8)

- SC1-2: Two tests with the sofa against the back wall, but with the doorway blocked, to determine the effect of room vitiation.

- BW1-7: Tests of two bookcases each, arrayed in a V shape, opposite the corridor doorway, with the door to the corridor open. (Figure 9)

- BP1-3: Similar bookcase tests, but with the rigid PVC sheeting product as an additional source of combustion products. (Figure 10)

- PW1-2: Tests of electric cable in the tray assembly, which was located parallel to the back wall of the burn room, with the doorway to the corridor open. (Figure 11)

Table 7. List of Room Fire Tests

Test #	Fuel	Location	Instruments	Notes
SQW1	8 cushions	Against rear wall	None	New drywall; paper burned off during this test
SQW2	8 cushions	Against rear wall	None	Did not go to flashover
BQW1	2 bookcases	Flat against rear wall	Hood only	Did not go to flashover
BQW2	2 bookcases	Against rear wall – V	Hood only	
SQW3	8 cushions	Against rear wall	Hood only	Did not go to flashover
SQW4	8 cushions	Against rear wall	Hood only	Did not go to flashover
SQM1	8 cushions	Mid-room	Hood only	Did not go to flashover
SQM2	8 cushions	Mid-room	Hood only	Did not go to flashover
BQW4	2 bookcases	Rear wall – V	None	
BW1	2 bookcases	Rear wall – V	Hood	Calcium silicate board replaced drywall
SW1	8 cushions	Against rear wall	All	Did not go to flashover
SW2	8 cushions	Against rear wall	All	Did not go to flashover
SW3	12 cushions	Against rear wall	All	Did not go to flashover
BW2	2 bookcases	Rear wall – V	All but FTIR	
BW3	2 bookcases	Rear wall – V	All but FTIR	
BW4	2 bookcases	Rear wall – V	All	
BW5	2 bookcases	Rear wall – V	All but FTIR	Did not go to flashover
BW6	2 bookcases	Rear wall – V	All but FTIR	
PQ1	Cable	Rear wall	All	
PQ2	Cable	Rear wall	All	
PW1	Cable	Rear wall	All	
PW2	Cable	Rear wall	All	
SW10	14 cushions	Rear wall	All	
SW11	14 cushions	Rear wall	All	
SW12	14 cushions	Rear wall	All	

Test #	Fuel	Location	Instruments	Notes
SW13	14 cushions	Rear wall	All	
SW14	14 cushions	Rear wall	All	
BP1	2 bookcases/ PVC	Bookcases: rear wall PVC sheet: room front	All	
BP2	2 bookcases/ PVC	Bookcases: rear wall PVC sheet: room front	All	
SC1	8 cushions	Rear wall	Location 1	Door closed; did not go to flashover
SC2	8 cushions	Rear wall	Location 1	Door closed; did not go to flashover
BW7	2 bookcases	Rear wall	All	
BP3	2 bookcases/ PVC	Bookcases: rear wall PVC sheet: room front	All	

Test Title Key [X(Y)Zn]

X: Fuel [S = sofas; B = bookcases; P = power cable]
Y: Q = heat release rate test
Z: M = combustibles located near middle of burn room (ceiling hole open)
   W = combustibles located near rear wall of burn room
   P = combustibles include PVC sheet
   C = doorway closed
n: test number for that set of combustibles and location

22

Figure 6. Photograph of 8-cushion Sofa

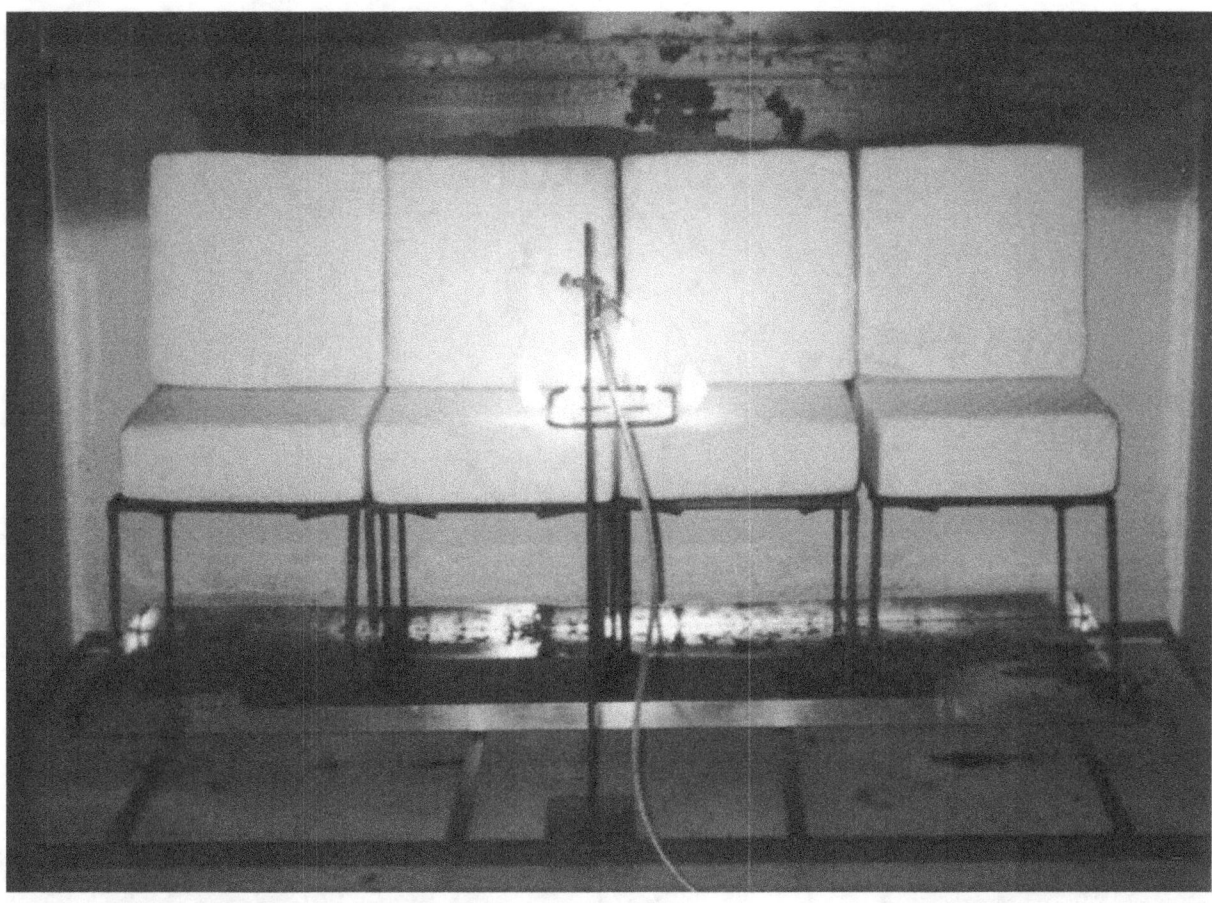

Figure 7. Photograph of 12-cushion Sofa

Figure 8. Photograph of 14-cushion Sofa

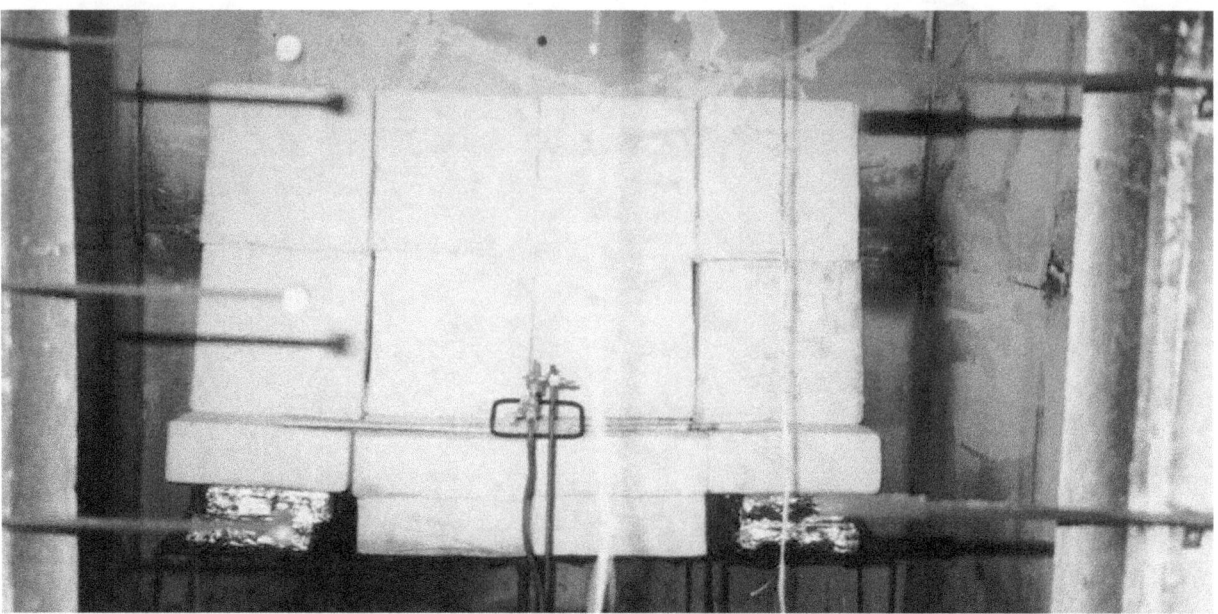

Figure 9. Photo of V-oriented Bookcases in the Burn Room

Figure 10. Photograph of Bookcases and PVC Sheet in the Burn Room

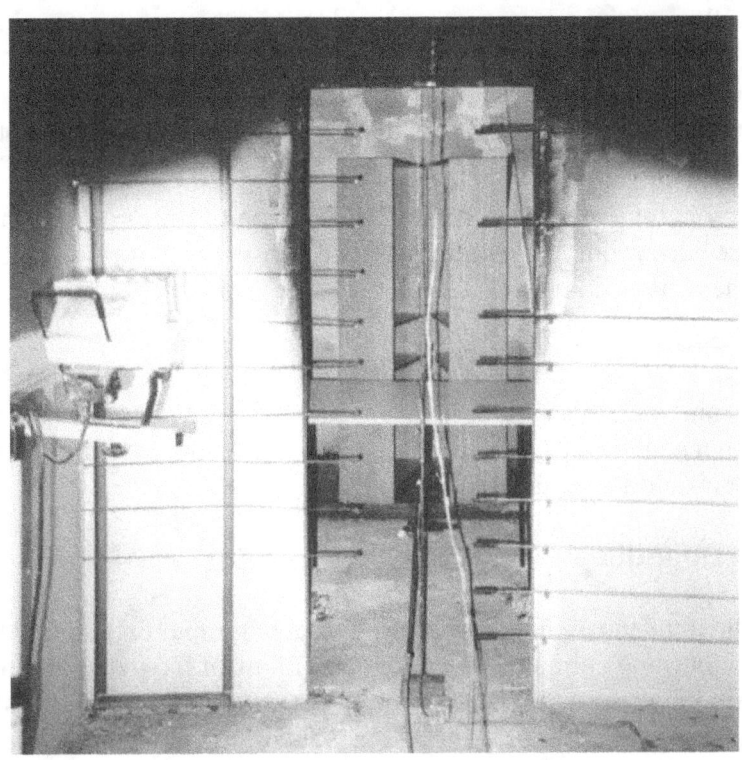

Figure 11. Photograph of Cable Trays in the Burn Room

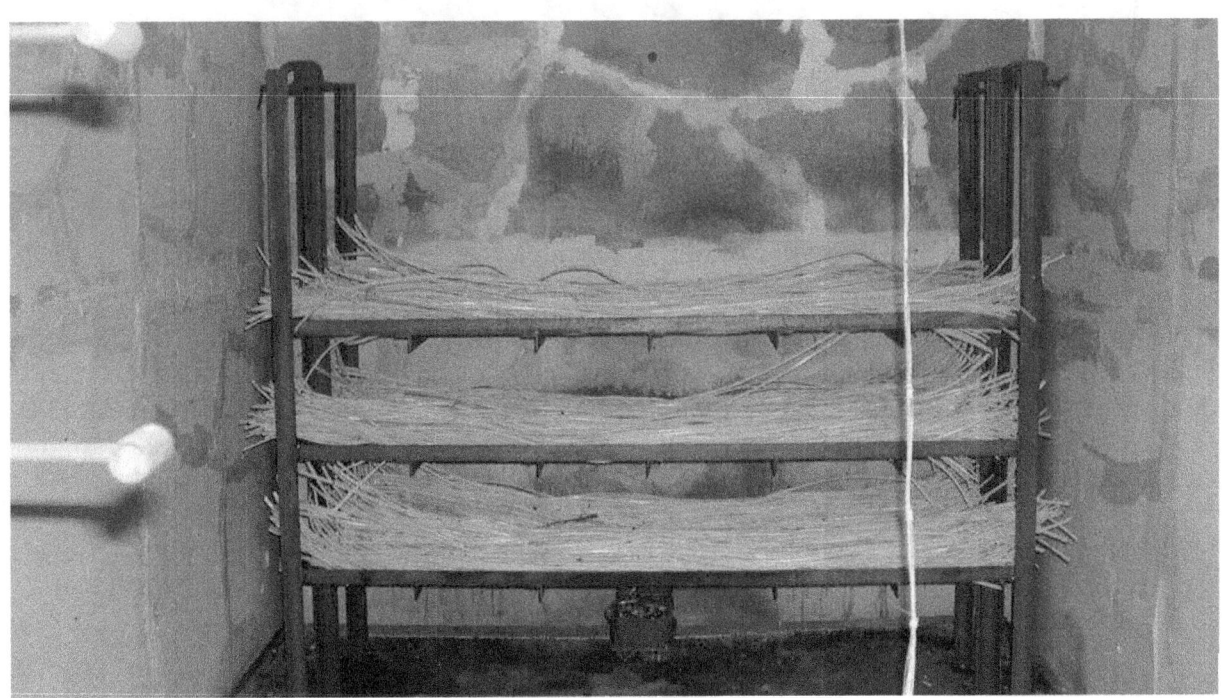

E. Measurements and Sampling Methods

Calculating the rate of heat release requires certain measurements in the exhaust duct: oxygen concentration, mass flow of the effluent stream, and temperature of the stream.

Determining the yields of the toxic gases requires both measurement of the concentrations of those gases and the other time varying factors that enable conversion of the concentrations into species yields:

- Mass flow through the doorway, which is a function of the door area through which the flow exits the burn room, the temperature and density of that flow, and the pressure differential across the doorway.

- Mass loss of each combustible.

The instrumentation and methods used to measure each of these quantities are discussed below. Readers interested in a general overview of large-scale fire testing data collection and analyses are referred to Peacock and Babrauskas.[14]

1. Exhaust Duct Quantities

Instrumentation in the 6 m square hood[10] was used to obtain input data for calculation of the rate of heat release of the burning combustibles. Concentrations of oxygen, carbon dioxide and carbon monoxide were made at a single point in the centerline of the exhaust stack. Temperature and pressure were measured at six positions and averaged to obtain single values for the calculation of the mass flow.

2. Temperature

Knowing the vertical temperature profile is central to:

- Defining the fraction of the doorway opening through which combustion products exited the burn room,

- Quantifying the exit flow from the burn room, and

- Characterizing the smoke flow down the corridor.

As part of the characterization of the flow through the doorway, a thermocouple tree was located in the doorway, approximately 100 mm (4 in) from the door edge. The 10 individual thermocouples were placed at heights of 0.53 m, 0.68 m, 0.83 m, 0.98 m, 1.13 m, 1.28 m, 1.43 m, 1.58 m, 1.73 m, and 1.88 m from the floor. The 10 thermocouples were an aspirated design characterized by Pitts et al.[15] and based on a design by Newman and Croce[16]. The shield had a diameter of 6.3 mm. The shield housed a type K chromel-alumel thermocouple constructed from 0.51 mm diameter wire. A flow of 18.9 L/min (at ambient temperature) was drawn through each aspirated thermocouple by a dedicated pump. The aspirated gases were filtered and dried before passing through the pumps.

Three additional trees of 12 bare-bead thermocouples were used to determine the vertical temperature stratification in other locations:

- in the burn room approximately 1 m from the doorway wall and 1 m from the adjacent side wall, and

- in the corridor 1 m from the room of fire origin and 1 m from the adjacent side wall.

These type K thermocouples, constructed from 0.25 mm diameter wire, were spaced evenly from floor to ceiling at 150 mm ± 10 mm (6 in ± 0.4 in) intervals, again, beginning 0.53 m from the floor.

A single type K thermocouple was located just below the centerline of the doorway lintel. This was used during the tests to assist in anticipating the onset of flashover.

In a past series of well-controlled gas burner tests[17], the standard uncertainty for peak gas temperature had been found to be ± 16 °C. This is expressed as the standard deviation of the peak values for 12 replicate tests. While random variation in the current experiments is expected to be comparable to these values, additional uncertainty due to variation in the ignition and fire growth of the fire sources from this test series can be expected. Replicate tests were conducted here to bound the uncertainty under the conditions of these tests.

3. Doorway Velocity

The other component of doorway velocity measurement was a vertical array of 10 bi-directional velocity probes designed for measuring the soot-laden doorway flows. These were based on a design developed by Heskestad (Figure 12).[18] Differential pressure from the two sides of the probes allows direct calculation of velocity and vent flow.[19] For the experiments in this report, vertical arrays of 10 bi-directional velocity probes and 10 corresponding aspirated thermocouples provide data for the calculation of vent outflow. Standard uncertainty in vent flow measurements has been reported to be approximately ± 10 %.[19] Replicate tests in the current series served to bound the uncertainty under the conditions of these tests.

4. Sample Delay Times

Delay times for gas flows from sampling locations within the test structure to the NDIR and FTIR gas analyzers were determined by introducing a pulse of gas into the sampling lines and measuring the time for each instrument to respond. Since the NDIR line temperatures were close to ambient during the tests and the FTIR lines were heated during these determinations, the values in Table 8 are very close to those experienced during the tests. Calculations for several of the experiments with changes in the delays ranging from 10 s to 40 s show less than 1 % change in calculated species yields.

Table 8. Instrument Delay Times and Standard Deviations (s)

Analyzer	Sampling Location			
	1	2	3	4
CO_2, CO, O_2	20 ± 2	34 ± 3	21 ± 1	9 ± 1
FTIR	3.0 ± 0.5	2.5 ± 0.5	--	3.0 ± 0.5

Figure 12. Schematic of Bi-directional Velocity Probes

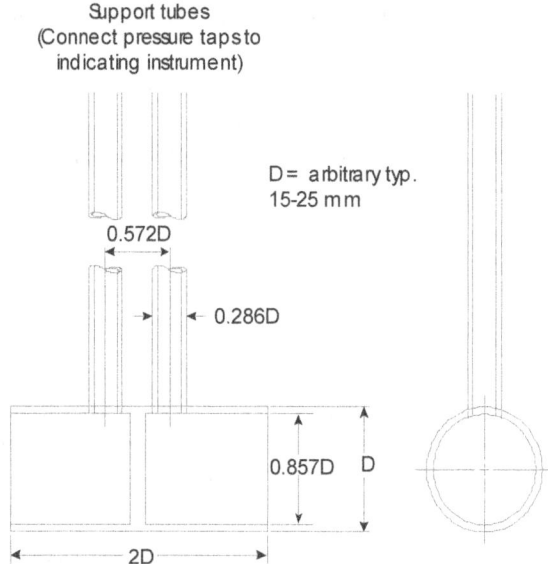

5. Sample Mass

During each test, the mass of the test combustible was recorded using one of two load cells.

- One with top surface measuring 1.5 m by 2.4 m (5 ft by 8 ft) was used to measure the mass loss of the initially ignited item (sofa, bookcases or cable trays). This load cell has a capacity of 1000 kg (2200 lb) with a rated resolution of 0.5 kg (1.1 lb).

- A second was used to measure the mass loss of the PVC sheet when it was used as a target object. The top surface of this load cell measured 0.6 m by 0.8 m (1.5 ft by 2.5 ft) and had a capacity of 150 kg (300 lb) with a resolution of 2 g (0.004 lb).

6. CO, CO₂, and O₂ Concentrations

As presented in Section II.B.3, samples for analysis of these gases were extracted from four locations. In the case of the four-probe arrays, this probe was the downstream (D) tube. Each probe was constructed of stainless steel tubing 12.7 mm (0.5 in) in outer diameter. The probes at the three corridor locations were inserted horizontally through the corridor sidewall and were thus roughly isothermally heated along their length to the layer height temperature at the time. The probe into the burn room was inserted through the ceiling. The sampling tips of the sampling probes were simply the blunt ends of the tubing. On the outside of the burn room or corridor wall, each probe D was connected to a length of unheated copper tubing 6.2 mm in outer diameter. Approximately 5 m downstream, this tubing was formed into a helical coil, which was immersed in an ice bath and then a dry ice bath to trap water vapor, aerosols and soot. A length of plastic tubing continued to the analyzers. A pump located on the downstream end of the train drew sample at an estimated rate of 10 L/min.

CO and CO_2 concentrations were measured continuously using nondispersive infrared (NDIR) analyzers. These instruments utilize absorption of infrared light at a single wavelength whose selection discriminates against absorption by strongly absorbing interferants such as water. None of the gases known to interfere with NDIR measurement of CO_2 and CO were expected to be present in sufficient quantity in these experiments to warrant correction to the absorption data. See Link *et al.*[20] for further details of the NDIR technique.

Oxygen measurements were made using paramagnetic analyzers. None of the other combustion gases known to have significant paramagnetic moments were expected to be present in sufficient quantity in these experiments to warrant correction to the data.

Prior to each test, flows of gas mixtures of known concentration insured correct operation of the analyzers and enabled any corrections to manufacturer-supplied calibration curves. Typically, these corrections were less than 1 % of the measured value.

For a series of well-controlled gas burner tests[17], the standard uncertainties for oxygen, carbon dioxide, and carbon monoxide concentrations were found to be ± 0.6 %, ± 0.4 %, and ± 0.06 %, respectively. These are expressed as the standard deviation of the peak values for 12 replicate tests. The random variations in the current experiments are expected to be comparable to these values. Additional variation (in these and all measurements) arising from variations in the ignition and growth rate of the fire are estimated in Section IV.F from the results of replicate tests.

7. Gas Sampling for FTIR and Ion Chromatographic Analyses

Samples for these analyses were extracted continuously at locations 2 and 4 (Section II.B.3). The sampling probes were the upstream (U) tubes in the four-probe arrays and were constructed of stainless steel tubing 12.7 mm (0.5 in) in outer diameter. At the outside of the corridor wall, each probe was connected to two polytetrafluoroethylene (PTFE) transfer lines 12.7 mm in outer diameter.

From each of the two probes, one line, *ca*. 8 m in length, went directly to one of two Fourier transform infrared (FTIR) spectrometers. These lines were heated to 170 °C tests to prevent the condensation of water, soot, and other nonvolatiles. There were no soot filters in the transfer lines since these also collect acid gases. While the acid gases can be extracted and analyzed after the test is over, one only obtains an integrated mass of each compound, and meeting the objective of this study requires time-resolved (at least pre- vs. post-flashover) concentration information.

The second lines, also *ca*. 8 m in length but unheated, were for transport of the gases for wet chemical analysis. Unlike the sampling for essentially continuous FTIR analysis, the ion chromatographic technique utilized batch samples, *i.e.*, samples accumulated over some time interval. Two samples were collected at each location in each experiment: one during the pre-flashover phase and one post-flashover.

The collection intervals differed with each experiment and were determined by the test coordinator. When the test coordinator determined that it was appropriate to begin sampling pre-flashover smoke, a valve was opened diverting the extracted flow from the heated PTFE tubing to, in order:

- a 250 mL glass impinger bottle containing 125 mL of 5 mM KOH in water conditioned to 18.2 MΩ –cm (ultra-low electrical conductivity);

- a 45 mm diameter PTFE filter (0.45 μm nominal porosity) to break up the gaseous aerosol and allow maximum collection in the first impinger.

- a 125 mL glass impinger bottle containing 100 ml of 5 mM KOH, also in water conditioned to 18.2 MΩ –cm;

- a Matheson 602 rotameter equipped with a control valve; and

- a pump.

At the end of the pre-flashover period, the flow was diverted to an exhaust line. When the test coordinator determined that it was appropriate to begin sampling post-flashover smoke, the flow was directed to a second identical set of impingers. At the end of this second collection period, the flow was again diverted to the exhaust line.

The flows through both analysis trains were measured prior to each test with an American Meter Company DTM-115 dry test meter. The flows to the FTIR spectrometers were maintained at 10 L/min ± 0.5 L/min using Matheson 602 rotameters with control valves. The flows to the impingers were similarly maintained at 1.00 L/min ± 0.055 L/min.

During the two closed room tests, the transfer lines from the probe at the downstream end of the corridor were moved to another probe located on the burn room centerline approximately 1 m inside the burn room door.

As with the fixed gas analyzers, the signal delay resulting from the combination of residence time in the sampling line and the FTIR spectrometer response time was determined using gas pulses introduced at each probe tip. The time delays were 2.5 s ± 0.5 s for the upstream measurements and 3.0 s ± 0.5 s for the downstream measurements.

8. FTIR Analysis

The infrared absorbance spectra of fire gases extracted from two locations inside of the test facility were measured simultaneously at a frequency resolution of 0.5 cm^{-1}. Photographs of the FTIR spectrometers used in making these measurements are shown in Figure 13. Both are Midac Illuminator spectrometers equipped with mercury cadmium telluride (MCT) detectors. The unit used to monitor the fire gases at location 2 is configured with a closed optical path through an internal (stainless steel) cell fitted with ZnSe windows. The other spectrometer, which was used predominately at location 4, but was moved to location 1 for two tests, is an open path unit consisting of separate source and detector modules. An external monel cell with KBr windows was positioned in the optical path between the two modules. Although both cells were nominally 10 cm long, their optical pathlengths, which were determined by fitting the measured spectrum of CO at a known concentration (216 µL/L) to calibration spectra, were significantly different. The pathlength of the stainless steel cell with ZnSe windows was found to be 8.2 cm, while the monel cell with KBr windows was 11.5 cm. The scanning rate of both spectrometers at 0.5 cm^{-1} resolution is approximately 1.5 spectra/s. The spectrometer at location 2 was programmed to signal- average over 2 spectra, whereas the spectrometer at location 4 was configured to signal average over 4 spectra. Therefore, the concentrations of the target compounds at locations 2 and 4 were updated every 1.3 and 2.7 seconds, respectively.

Figure 13. Photograph of the FTIR Spectrometers

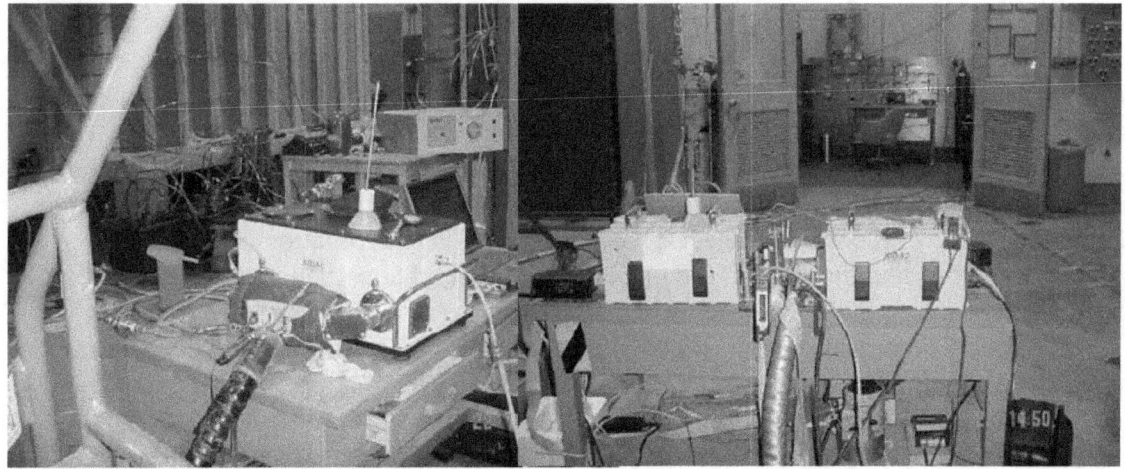

In selecting cell sizes for these spectrometers, there were two conflicting considerations. A larger cell with a multi-pass optical path (*ca.* 1 m) offers higher sensitivity, a benefit when examining trace compounds, but a potential saturation problem with the major species. A smaller cell with a smaller volume offers better time resolution, an important issue when the combustion conditions in the fire are likely to be changing during a test. We deemed the latter consideration to be more important and used the smaller optical cells.

A typical spectrum measured by FTIR spectroscopy during the fire tests is displayed in Figure 14. The series of peaks extending from about 3050 cm^{-1} to 2600 cm^{-1} are due to HCl. In this case, it is possible to resolve the individual frequencies corresponding to changes in the population of rotational states as the H-Cl bonds vibrate. This is usually only possible for small gas phase molecules. There are three spectral features due to CO_2 that are evident in this spectrum. The most intense, centered at 2350 cm^{-1}, corresponds to asymmetric stretching of the two C=O bonds. The symmetric stretch is not observed because there is no change in dipole moment when both O atoms move in phase. The second feature, seen as two distinct peaks centered at about 3650 cm^{-1}, is an overtone band that derives from the simultaneous excitation of these bond-stretching modes. The third peak at about 650 cm^{-1}, only partially visible in Figure 14, is due to the out of plane bending of the molecule. Also shown are the bands due to the C≡O and H-CN stretching vibrations of CO and HCN centered at about 2150 cm^{-1} and 3250 cm^{-1}, respectively. The latter band interferes strongly with the C-H stretching vibration of acetylene, which is also present in the fire atmospheres. The remaining peaks in this spectrum are due to H_2O.

As noted earlier, the toxicants under investigation were: CO_2, CO, HCN, HCl, HF, HBr, NO, NO_2, and CH_2=CH-CH=O (acrolein). The relative concentrations of these compounds were determined from IR absorbance measurements of the fire gases using Autoquant 3.11. This is a software package for performing real time quantitative analyses of target compounds, which is based on the Classical Least Squares (CLS) algorithm as described by Haaland *et al.*[21] In this method, the measured spectra are fit to linear combinations of reference spectra corresponding to the target compounds.

The implementation of this method requires that the user supply calibration spectra corresponding to the IR absorbencies of known concentrations of the target compounds. These data were obtained from a quantitative spectral library assembled by Midac[22] and from a collection of spectra provided by Federal Aviation Administration staff who had performed bench-scale fire tests on similar materials.[23] In our implementation, the least squares fits were restricted to characteristic frequency regions or windows that were selected in such a way as to maximize the discrimination of the compounds of interest from other components present in fire atmospheres. For some of the gases, a piece-wise concentration interpolation (PWCI) was employed to correct for deviations from the linear dependence of absorbance on concentration. This procedure, which requires spectra of the calibration gases at multiple concentrations, involves interpolating between independent CLS analyses such that the predicted concentrations fall between the concentrations of the calibration mixtures.[24]

The identities of the target compounds (as well as other compounds that absorb at the same frequencies and must, therefore, be included in the analyses), their corresponding concentrations (expressed in units of ppm for a mixture of the calibration gas and N_2 in a 1 meter cell), and the characteristic spectral windows used in the quantitative analyses are listed in Table 9.

Figure 14. Representative Spectrum of the Fire Gases Extracted during a Test

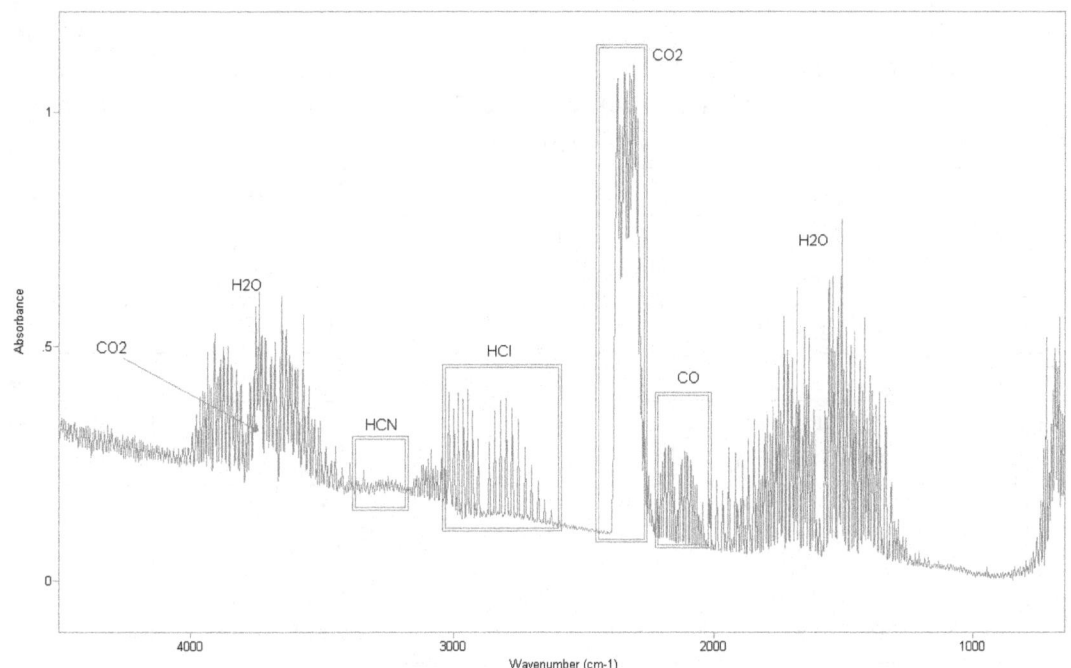

Also listed in this Table are minimum detection limits (MDLs) for each of the target compounds. These values, which represent the lowest concentrations that can be measured with the instrumentation employed in these tests, were estimated as follows. The calibration spectra were added to test spectra (which, when possible, were selected in such a way that only the compound of interest was not present) with varying coefficients until the characteristic peaks of the target compounds were just discernible above the baseline noise. The value of signal averaging over *ca.* 100 spectra was included. The MDL values reported in Table 9 were obtained by multiplying these coefficients by the known concentrations of the target compound in the calibration mixtures.

Water, methane and acetylene are included in the quantitative analyses because they have spectral features that interfere with the target compounds. The nitrogen oxides absorb in the middle of the water band that extends from about 1200 cm^{-1}-2050 cm^{-1}. Consequently, the limits of detection for these compounds are an order of magnitude higher than for any of the other target compounds. Thus, it is not surprising that their presence was never detected in any of the tests.

Table 9. Calibration Spectra for FTIR Spectroscopy

Compound	Concentration (vol. fraction-m x 10^6)	Temperature (K) (P = 101.3 kPa)	Frequency Windows (cm^{-1})	Minimum Detection Limit (vol. fraction x 10^6)
C_2H_2	39	170	3190-3420	-
C_3H_4O	225	100	850-1200 2600-2900	10
CH_4	48, 422	170	2800-3215	
COH_2		100	2725-3000	50
CO	51, 241, 1460, 8977, 17650	170	2010-2250	10
CO_2	4785, 9125, 16329	170	715-724 2250-2400	5
H_2O	10000	170	1225-2150 3400-4000	-
HBr	226	170	2400-2800	50
HCN	51, 115	170	710-722 3200-3310	15
HCl	987	170	2600-3100	15
HF	2025	170	4000-4150	5
NO	512	121	1870-1950	500
NO_2	77	121	1550-1620	100

9. Acid Gas Analysis by Ion Chromatography

After the samples were collected, the contents of each impinger bottle was transferred to a tared plastic bottle and weighed. The PTFE filters were removed from their holders and placed in the plastic bottle containing the contents from the first impinger. Approximately 3 mL aliquots were removed from each plastic bottle and filtered through an 0.45 μm filter (to remove any particulate matter) into an ion chromatography vial for analysis.

A Dionex DX600 ion chromatograph equipped with a GP50 gradient pump and a CD25 conductivity detector and IonPac AS11 column was used to analyze for Cl^-, Br^-, NO_2^-, and NO_3^-. The instrument was calibrated as follows:

1. Stock solutions of Cl^-, Br^-, NO_2^-, and NO_3^-, nominally 1000 mg/L each, were prepared by dissolving the appropriate amount of salt in 18.2 MΩ-cm water.

2. Using serial dilutions of these solutions, curves of anion peak height vs. anion concentration were determined.

3. A calibration function was derived from these data. Least squares analysis of the resulting lines indicated a high degree of linearity (r^2 = 0.9915) over the anion concentration range 1.0 mg/L to 5.0 mg/L.

Comparisons of the values with the corresponding results obtained from the FTIR analyses indicated that the wet chemical method gave far lower yields (by more than an order of magnitude) of HCl and HCN. Experiments performed at the conclusion of the test matrix revealed that significant amounts of impinger water were being drawn through the unheated lines connected to the impinger bottles during the tests. Based on this observation, we have concluded that most of the gases of interest must have dissolved in this water, which presumably clung to the walls of the PTFE lines in form of droplets and was, therefore, never accounted for in the wet chemical analyses. Thus, no further discussion of these results is included in this report.

10. Smoke Mass

Of interest to this project is the extent to which reactive gases diminish in concentration as they travel from the fire vicinity. One mechanism for the loss of these gases is by adsorption on particulate matter. Accordingly, the concentration of smoke was determined at locations 2 and 4 in the corridor where the reactive gas concentrations were also measured.

As with the impinger bottle samples, two samples were collected at each location in each experiment: one during the pre-flashover phase (Probe T) and one post-flashover (Probe B). The specific sampling time intervals were nominally the same as for the impinger bottle sampling.

At the outside of the corridor wall, each probe passed through a 42 cm long water jacket heated to 55 °C to prevent the condensation of water and other volatiles. The smoke was collected on a polytetrafluoroethylene (PTFE) filter with a 2 μm pore size housed in a stainless steel filter holder also heated to 55 °C. The collection efficiency for this filter was at least 96 % for particle sizes of 0.035 μm and larger. This size range includes essentially all the smoke particles. The flow through the filter was 50 cm^3/s (3 L/min).

Because a significant mass of sampled smoke might collect on the interior wall of the sampling line (before reaching the filter), a cotton pad was used to clean the inside of the tube in a manner similar to cleaning a gun barrel. The fraction of smoke deposited in the tube varied from about 15 % for the bookcase fires to as large as 75 % for the tests involving bookcases plus a PVC sheet. Even with this cleaning procedure, some of the smoke was retained on the tube walls; this was estimated to be no more than 15 % of the total.

Repeat weighings of the collected smoke were made after storing the filters overnight to assess the impact of condensables. The change in mass was typically less than 2 %. The moisture effect is typically small for PTFE filters; however, it can be a substantial effect for quartz or fiberglass filters. The typical uncertainty in the filter weighing was about 0.05 mg for a 1 mg filter weight. The filter weights ranged from 0.4 mg to 40 mg. The mass concentration of smoke corrected to 101 kPa and 25 °C ranged from 0.2 g/m^3 to 20 g/m^3.

11. Heat flux

As one measure of whether and when flashover was achieved in the burn room, the radiant flux at the floor in the burn room enclosure was measured using a Medtherm 20 W/cm^2 Gardon-type total heat flux gauge. The manufacturer's specified accuracy was ± 3 % with a repeatability of ± 0.5 %. The gauge was positioned looking upward, with its measuring surface flush with the floor surface of the burn room, centered between the two side walls at approximately 1 m (3.3 in) from the doorway.

12. Video

The progress of most tests was monitored using two Super 8 video cameras. One was located beyond the open end of the corridor and viewed the full length of the corridor and the interior of the burn room. The second was located in the forward lower corner of the burn room and was directed upward toward the burning fuel. For the two tests where the sofas were located in the burn room center facing toward the rear, this camera was relocated to the lower rear corner. Only one camera was active during the tests with the burn room door closed. This was located in the lower front corner of the burn room as above.

13. Additional Data

The building temperature, pressure and relative humidity were recorded at the beginning of each test.

F. Data Collection

1. Hardware

The signals from the various measurement devices (except the FTIR spectrometers) were collected using the NIST Large Fire Facility data acquisition system (DAQ). The DAQ is currently configured to collect data both from instruments in the exhaust duct of the 6 m hood for calorimetry calculations and from any other instruments or sensors located in the vicinity of the experiment.

The hardware consists of National Instruments components. The primary DAQ board is unit PCI6052E, which has a PCI bus computer interface, 16 analog channels, 16 bit resolution, 333 ksamples/s sampling rate, eight digital I/O channels, and two 24-bit counter/timers. An SCXI1001 chassis box holds seven modules for multiplexing channels onto one DAQ board channel. The SCXI1102B module was the primary component used for multiplexing and signal conditioning. Each module has 32 channels capacity, a 200 Hz low pass filter, and 100X gain option for low voltage signals. Each module was connected to a TC2095 shielded terminal block, which collects signals of up to 10 V amplitude and provides cold junction compensation for any thermocouples voltages.

A total of 130 channels were scanned for these experiments; 40 were inputs to the calorimetry calculations, and the remaining 90 were unique to this series of experiments. While the DAQ has the capability to scan each channel 300 times per second, the typical rate in these tests was 200 scans/s. The 200 values for each channel were averaged for an overall output and storage rate of 1 sample/s. In addition, during each 1 s cycle or loop, numerous calculations were performed "on the fly" and the output stored in a second data file.

2. Event Marking

Three channel markers were used to keep track of specific events during a fire test:

- Event Marker 1 was used to log the times for igniting and extinguishing the pilot burner and extinguishing the fire at the end of the test.

- Event Marker 2 was used to log the times for initiating and ending the pre-flashover sampling period.

- Event Marker 3 was used to log the times for the occurrence of flashover and the beginning and ending of the post-flashover sampling period.

Event marking put simulated voltage spikes from 0 to 5, 10 to 15, or 20 to 25 for one second in the channels set up for the markers. Plotting the data in these channels along with, *e.g.*, concentration and mass loss data enabled relating the evolving combustion product chemistry to the changing fire phenomenology in the test.

3. On-the-fly conversions

All data were stored as voltage magnitudes without modification in a "raw" data file. The exception was that all temperature measurements were converted in real time to degrees Celsius and stored as such. A second file was created simultaneously. For this file, all the other signals were converted in real time to the physical quantities they represented: CO, CO_2, and O_2 dry volume fractions, pressures, heat fluxes, smoke attenuations, and load cell masses. These were in turn combined to also generate flow velocities, duct volume flow, duct mass flow, and heat release rate. Up to 6 levels or steps of conversion and combination were used to complete all the calculations. For heat release calorimetry testing, data were combined after estimating the relative delays for each instrument and synchronizing the data generated by a given moment of the fire. For the remainder of the tests, no delays were implemented and the relative time responses of the measurements were post-processed.

4. Real time presentation

Prior to this research project, all channels of data were available digitally in either raw or converted form, and two channels could be user-selected for graphical display. Also, calorimetry analyzer readings, velocities, and heat release rate were displayed graphically with up to 6 min of history.

For this project, several measurements were deemed important for periodic monitoring during each test and thus additional graphical presentations were created. These graphs were: all CO, CO_2, and O_2 volume fractions from the NDIR and paramagnetic analyzers, the temperatures from the thermocouple tree in the fire room, the doorway temperatures from the aspirated thermocouples, the velocities in the doorway, and the four temperatures from the soot probes. The fire room temperature graph could at any time be changed to plot any group of 12 adjacent channels.

5. Storage for analysis

Each scan and calculation loop included a file writing procedure as well. Three files were generated during a test with suffixes "raw," "adj," and "ZS:"

- The "raw" file recorded raw voltage values only, plus the addition of the 3 marker channel values.

- The "adj" file recorded real channels with their simple one-step conversions applied and all of the artificial channels that combined real channels in one or more steps. Also, if the user specified, the "adj" file listed results of calculations that were combined so as to synchronize the inputs to the same event. Thus, in such a time-adjusted "adj" file, *e.g.*, a burst of flame in the fire room would line up in time with all the following: the increase in specimen mass loss, an increase in temperature 20 m down the duct, and the corresponding increase in CO_2 sampled 20 m down the duct and then transported 30 m into the analyzer with its own inherent processing delay.

- The third file, with the suffix "ZS," listed the voltages generated by the zeroing and spanning processes for the gas analyzers as well as the calibration signals for the response of pressure transducers, load cells, etc. The former set are linear and interpolation between the two calibration points enables assignment of concentrations to the recorded voltages. The instruments in the latter group are generally non-linear or nearly linear, and these have been pre-calibrated over the necessary range. These calibration equations require a single point each to enable conversions from voltage to, *e.g.*, Pa or kg.

G. Test Procedure

There were a number of steps taken before the ignition of the test specimen:

- Setting the test specimen in place and recording its precise location and orientation;

- Describing the test specimen (*e.g.*, the number of cushions in a sofa, their relative positions, the wiring of the vertical cushions to the metal frame, recording the mass of the specimen);

- Ensuring the ignition device was properly placed;

- Inspecting the following for test readiness:
    - back-up suppression capability (standpipe and fire hose),
    - firefighters in turn-out gear,
    - gas racks zeroed and spanned,
    - video cameras running,

- cold traps filled with ice/dry ice,
- gas rack identification numbers recorded,
- aspirated thermocouples running,
- pumps operational for smoke collection lines and wet chemistry sampling lines,
- FTIR spectrometers calibrated,
- test and DAQ clocks synchronized
- bi-directional probes operational,
- heat flux gauge cleaned,
- and "flashover indicator" (balled-up piece of newspaper) placed on the floor of the burn room in view of the water-cooled camera,
- ambient pressure, temperature, and humidity recorded.

Next, two minutes of "background" values were recorded for all instruments. This was followed by a 10 s countdown to ignition during which all of the separate apparatus functions (data acquisition system, FTIR spectrometers, wet chemistry sampling, and smoke sampling) were further synchronized.

A firefighter, located in the burn room, ignited a portable gas torch. The propane flow to the ignition burner(s) was initiated, and the firefighter lit the burner. This time was recorded.

For the sofa and cable tray fires, the burner was kept lit until the test manager determined that the test specimen was ignited. This was defined as sustained flaming over a volume distinctly greater than that of the burner and was followed by direct observation through the doorway and by monitoring the output of the in-room video camera. The burner was then turned off and the firefighter withdrew, removing the burner from the burn room in the case of the sofa fires. [The sand burners used in the cable tray fires were not removable.] Both the times of ignition and the time of burner extinguishment were recorded.

For the sofa fires with the room doorway closed, the test manager followed the ignition process from the in-room video output. The electric match was not moved following ignition.

For the bookcases, it was difficult to determine when the particleboard was able to sustain burning. Following an apparent ignition, the flames would occasionally recede. In those cases, the burner was again ignited until significant flaming was observed, at which point the burner was extinguished. The time of assured ignition was recorded.

When the fire effluent was actively flowing out of the room of origin, sampling of the pre-flashover smoke to the various instruments was initiated and the time recorded. The target time interval for pre-flashover measurements was three minutes, although this time was frequently shortened when the test manager estimated that flashover was imminent. The time at which pre-flashover analysis stopped was recorded.

The time of flashover was recorded when the test manager observed flames extending from the room of origin into the hallway. This observation generally coincided with flaming of the paper placed on the floor in the room of origin. [Occasionally, incoming air from the hallway to the lower layer of the burn room blew the paper to a location where it did not see the high radiant flux from the upper layer.]

Post-flashover measurements were initiated shortly after flashover. Again, the desired measurement period was 3 minutes, although some tests became fuel-limited and reverted to pre-flashover burning conditions, necessitating early measurement termination. Both the start and end of this period were recorded.

At this point, one or more firefighters proceeded to extinguish the fire. For low level flaming, a hand-held fire extinguisher could be used. More often, the firefighter applied a stream of water from a garden hose. For large fires, they could have used a 1½" standard hose line water stream, but this was not necessary in these experiments. The times at which fire suppression started and the time at which the flames were quenched were recorded.

The closed-door tests self-extinguished, presumably due to a lack of oxygen. Once the test measurements were completed, the firefighters opened the door, which led to reignition. The small fire was then suppressed with a hand-held fire extinguisher.

III. CALCULATION METHODS

The following nomenclature is used in this section:

A	Cross sectional area of the exhaust duct (m^2)
C	Orifice plate coefficient ($kg^{1/2}\ m^{1/2}\ K^{1/2}$)
E	Net heat release per unit mass of oxygen consumed ($kJ\ kg^{-1}$)
F	Carbon mass fraction in the fuel
$\dot{m}$	Mass flow for gases or mass loss rate for combustible material ($kg\ s^{-1}$)
M	Molecular weight ($kg\ kmol^{-1}$)
p	Pressure (Pa)
$\dot{q}$	Heat release rate (kW)
T	Temperature (K)
W	Vent width (m)
X	volume fraction (dimensionless)
y	species yield (dimensionless)
α	Combustion expansion factor (dimensionless)
$\Delta\varphi$	volume fraction
ε	combustion efficiency
ρ	gas density ($kg\ m^{-3}$)
φ	Oxygen depletion factor (dimensionless); see below for definition.

Superscripts

A	refers to concentrations in the analyzer
0	refers to initial conditions for incoming duct gases

Subscripts

N	neutral plane
amb	ambient air
dry	Dry air
$duct$	incoming duct gases
f	Fire
g	refers to a specific gas of interest, *e.g.*, CO, CO_2
not	refers to the notional yield of a combustion product
s	smoke
t	top of the vent, total
v	vent

A. Heat Release Rate

The peak rate at which a product generates heat is the most fundamental fire property for use in fire hazard analysis. As such, the following text documents how the values were calculated for the "Q" tests, *i.e.*, those in which the ceiling hole was open. The results were not used in calculating the yield information that is the primary focus of this project.

The rates of heat release were calculated using the oxygen consumption principle, *i.e.*, the heat released during complete combustion of a specimen is 420 kJ per mole of O_2 consumed, with only a ±5 % dependence on the material(s) combusted.[25] This changes only slightly for the degree of incompleteness of combustion experienced in fire tests.[26] If all of the fuel carbon were to be partially combusted to CO rather than fully combusted to CO_2, the constant is reduced by about 22 %. The generation of carbonaceous material (soot) acts in the opposite direction, with a 10 volume % generation of carbon increasing the constant by less than 3 %. In this series of tests, the [CO]/[CO_2] ratio did not exceed 0.1 and the soot fraction was of the order of 1 % to 10 %. Thus, we corrected for the actual CO fraction and neglected the effect of the formation of soot and other products of incomplete combustion.

Since in each of the "Q" tests, all the effluent from the fire was collected in the exhaust hood, the total rate of heat release from the room can be determined using Janssens' formulation:[27]

$$\dot{q} = \left(E\phi - (E_{CO} - E)\frac{1-\phi}{2}\frac{X_{CO}^A}{X_{O_2}^A} \right)\frac{M_{O_2}}{M_{duct}}\dot{m}_{duct}\left(1 - X_{H_2O}^0\right)X_{O_2}^{A^0}$$

where:

$$M_e = \left(1 - X_{H_2O}\right)\left(X_{O_2} + 4X_{CO_2} + 2.5\right)4 + 18$$

$$\dot{m}_e = C\sqrt{\frac{M_{dry}\Delta p}{M_e T_e}}$$

$$\frac{\dot{m}_{duct}}{M_{duct}} = \frac{\dot{m}_e}{M_e}\left(\frac{\left(1 - X_{H_2O}\right)\left(1 - X_{O_2}^A - X_{CO_2}^A - X_{CO}^A\right)}{\left(1 - X_{H_2O}^0\right)\left(1 - X_{O_2}^{A^0} - X_{CO_2}^{A^0}\right)}\right)$$

$$\phi = \frac{X_{O_2}^{A^0}\left(1 - X_{CO_2}^A - X_{CO}^A\right) - X_{O_2}^A\left(1 - X_{CO_2}^{A^0}\right)}{X_{O_2}^{A^0}\left(1 - X_{O_2}^A - X_{CO_2}^A - X_{CO}^A\right)}$$

Simplifications are available, with some loss of precision, if concentrations of some of the gas species are not measured. For instance, the concentration of water vapor in the exhaust duct flow was not measured and was assumed to be the same as that measured in the ambient air due to the high ambient air dilution of the room effluent in the exhaust duct.

The overall uncertainty in peak heat release determinations results from uncertainties in the measured terms in the above equations and from variability in the fires themselves. The former can be high, especially for smaller fires due to the small difference between concentrations of oxygen in the ambient air and the combustion-vitiated air. In one study[28], coefficients of variation ranged from 4 % to 52 % for a wide range of experiments. The worst case resulted from an oxygen depletion of only 0.26 % in a 100 kW fire. Since the peak rates of heat release in the current series of tests are in the range of 1 MW, we estimated that the uncertainty in the peak rate of heat release was about ± 10 %, including the aforementioned uncertainty in the value of E.

The overall uncertainty is typically estimated from replicate tests, such as were performed here for the sofas. The repeatability of y_{CO2} was ± 23 %. Most of the carbon was fully oxidized to CO_2 and the heat release per mole of oxygen consumed is not very sensitive to the degree of completeness of combustion. Thus, this is a fair indicator of the overall uncertainty. This is consistent with a prior study, in which Babrauskas[29] found the precision for the peak rate of heat release from upholstered furniture fires, both in the open and enclosed in a room, to be within 15 % for fires of 2.5 MW.

B. Mass Loss Rate

The mass loss rates were determined from the slopes of the plots of mass vs. time. The noise in the instantaneous measurements was reduced using a running seven- point linear regression of the mass loss data. The slope at the midpoint of the seven-point interval was taken to be the mass loss rate at the midpoint. The measurement uncertainty is derived from the linearity and sensitivity limit of the load cell.

C. Combustion Efficiency

The combustion efficiency was estimated from the fractions of the fuel carbon that were oxidized to CO and CO_2:

$$\varepsilon = \quad X_{CO2}/(X_{CO2} + X_{CO})$$

The small contributions of pyrolyzed and partially oxidized combustion products are ignored.

The uncertainty in the combustion efficiency is derived from the uncertainties in uncertainty in the measured volume fractions (or the yields) of CO and CO_2. (Table 14).

D. Doorway Flows

Computation of mass flows through openings uses the velocity and density profiles in the opening:[30]

$$\dot{m}_v = \int_{h_N}^{h_t} \rho v W dv = \int_{h_N}^{h_t} \rho_{amb} \frac{T_{amb}}{T_v} v W dv$$

Doorway profiles of the velocity, v, were measured with a series of bi-directional probes from the design of Heskestad.[18] Corresponding temperature measurements and the ideal gas law provided the densities, ρ. A corollary of the overall mass flow calculation was used to calculate the flow of individual gases with additional measurement of the gas concentration as:

$$\dot{m}_g = \dot{m}_v \left(x_g \frac{M_g}{M_v} \right)$$

We then used an average value over each sampling period for the yield calculations.

The uncertainty in these flow measurements is not small. Since the pressure drop across an opening passes through zero as the flow changes direction at the height of the neutral plane, measurement of the velocity profile in a doorway is particularly difficult. Estimation of the pressure in the extreme lower resolution of the instrumentation (as the pressure drop approaches zero) adds to the uncertainty of the measurement. For the same range of experiments noted above for heat release rate, the repeatability of the vent mass flow calculation averaged 35 %.[28] The measurement uncertainty in the current experiments is expected to be comparable to this.

The overall uncertainty in the doorway flow measurements also includes the effect of variation in the ignition and fire growth. The overall uncertainty was estimated using data from replicate tests.

E. Global Equivalence Ratio

The fuel to air global equivalence ratio is defined as:

$$\frac{\dot{m}_{\text{fuel}} / \dot{m}_{\text{air}}}{[\dot{m}_{\text{fuel}} / \dot{m}_{\text{air}}]_{\text{stoichiometric}}}$$

The terms in the numerator are the mass loss rate and the doorway flow. The denominator is calculated from the empirical formula of each fuel from Section II.C. The uncertainty in the equivalence ratio is the sum of the uncertainties in the mass loss rate, the doorway flow and the empirical formula.

F. Notional Gas Yields

The notional, or maximum possible, gas yields (Table 10) were calculated as follows:

- CO_2: Assume all the carbon in the test specimen is converted to CO_2. Multiply the mass fraction of C in the test specimen (Table 2 or Section II.C) by the ratio of the molecular mass of CO_2 to the atomic mass of carbon.

44

- CO: Assume all the carbon in the test specimen is converted to CO. Multiply the mass fraction of C by the ratio of the molecular mass of CO to the atomic mass of carbon.

- HCN: Assume all the nitrogen in the test specimen is converted to HCN. Multiply the mass fraction of N by the ratio of the molecular mass of HCN to the atomic mass of nitrogen.

- HCl: Assume all the chlorine in the test specimen is converted to HCl. Multiply the mass fraction of Cl by the ratio of the molecular mass of HCl to the atomic mass of chlorine.

Table 10. Estimated Notional Yields of Toxic Products (mass fraction)

	Bookcase	Sofa	PVC Sheet	Cable
CO_2	1.72	2.00	1.55	2.11
CO	1.09	1.27	0.98	1.33
HCN	0.057	0.193	-----	0.040
HCl	0.0026	0.0070	0.537	0.332

The uncertainty in the notional yield values is determined by the uncertainty in the prevalence of the central element (in the bullets just above) in the combustible. These uncertainties were estimated in Section II.C.

F. Measured Gas Yields

For the open-door tests, the yields of the gases were determined using the following steps:

- Define the pre- and post-flashover time intervals.

- Calculate the instantaneous mass flow of each gas from the room during those two intervals using the calculated instantaneous total mass flow through the doorway and the concentration measurements at location 2.

- Average the resulting mass flow of each gas over each of the two time intervals.

- Determine the average test specimen mass loss rate during each of the two time intervals.

- Calculate the ratios of the average gas mass flows to the average mass loss rate during each of the two intervals.

For the closed-room tests:

- Calculate the position of the upper layer interface as the height where the measured temperature has risen significantly above ambient. For these calculations, this was taken as a rise of at least 20 % of the temperature rise from the floor to the ceiling.

- Assume that the temperature of the upper layer can be represented by the average top-to-bottom temperature as measured by the thermocouples in the tree in the burn room.

- Assume that the upper layer is defined by the height at which the temperature was 20 % of the range defined by the highest and lowest thermocouples.[31]

- Assume the upper layer is well mixed.

- Use the measured volume fraction of the individual gases (location 1) and the ideal gas law to calculate the mass of each species in the upper layer.

- Calculate the ratio of these masses to the specimen mass loss, as a function of time.

- Use the final time-dependent yield as an indicator of the average species yield during the test, or (if there is marked variation with time) segment the test into separate yield values.

For the PVC sheets:

- Since the mass loss was negligible before flashover, only post-flashover results are determined.
- Since the Cl content of the bookcases is very small, assume all the HCl is from the PVC sheet.
- Since the N content of the PVC sheet is very small, assume all the HCN comes from the bookcase.
- Since the scatter in the CO and CO_2 yields is comparable to any differences between tests with and without the PVC sheet, yield values from the PVC sheet for these two gases are not calculable.

The uncertainty in the yield values results from:
- the uncertainty in the species concentration measurement,

- the sensitivity of the yield to the selected time interval (determined by varying the length and timing of the interval),

- the uncertainty in the specimen mass loss,

- the uncertainty in the species mass flow out the doorway (for open door tests), and

- the quality of the assumptions inherent in the calculation of the mass of product in the upper layer (for closed room tests).

For the closed room tests, the uncertainty can be further estimated by comparing the yield values from the early combustion with those from the pre-flashover segments of the open door sofa tests.

G. Smoke Yields

The smoke yield is determined by the carbon balance method. This method requires a determination of the ratio of the smoke mass in a given volume to the total mass of carbon in the form of gas or particulate in the same volume. This is accomplished by dividing the smoke mass collected on a filter by the sum of the smoke mass and the mass of carbon contained in the forms of CO and CO_2. The equation for calculating smoke yield, y_s, as expressed in terms of CO_2 and CO concentrations, is given by:

$$y_s = \frac{f \, m_s}{[\, m_s + 12n_t \, (\Delta\phi(CO) + \Delta\varphi(CO_2)]}.$$

The quantity f is the carbon mass fraction of the fuel, m_s is the mass of the smoke sample collected on a filter, n_t is the number of moles of air sampled, and the constant 12 represents the molar mass of carbon in grams. The quantities $\Delta\varphi(CO)$ and $\Delta\varphi(CO_2)$ are the volume fractions of CO and CO_2 of the gas sample taken during the test minus the ambient background concentrations of these gases. In this equation, the other carbon containing gases are neglected. This is a good approximation for overventilated burning; however, for underventilated burning the concentration of unburned hydrocarbons could be appreciable. Leonard et al. [32] found that these three species accounted for 82 % of the carbon for underventilated diffusion burning of both methane and ethylene at a global equivalence ratio of 1.32 and values in the range of 72 % to 74 % at an equivalence ratio of 1.52. This indicates that for the post flashover measurements, the smoke yields may be overestimates. There is one other approximation in using the equation above for computing y_s. The denominator is the sum of the carbon masses from the smoke, CO_2, and CO. The carbon fraction of the smoke mass m_s is at least 95 % carbon. This implies that the m_s term in the denominator has been overestimated by as much as 5 %. However, over 80 % of the carbon produced by the combustion is in the form of CO_2 and the maximum 5 % overestimate of the carbon mass corresponds to a 1 % overestimate of the total carbon mass. Thus, the fact that the carbon content of smoke is 95 % or greater carbon has at most a 1 % affect on the value of Y_s

IV. RESULTS

In the following discussion and tables, the analysis will not include some data for the following reasons:

- In the early, scoping tests there were no combustion product measurements taken.

- An instrument may have failed during a given test. In such a case, we attempted to use results from similar tests to avoid having to discard the entire test. These cases are noted in the text and tables that follow.

- Due to the low burning rates during the early segments of some tests, the gas sampling probes were not clearly in the upper layer of the corridor, as indicated by the temperature profile from the thermocouple trees. This was especially true for pre-flashover measurements at Location 4. In these cases, the data were discarded.

- The data from Location 3 were determined to add little to the results.

- We don't use the data from Location 1 except in the closed room tests. It is conceptually possible to use the location 1 CO and CO_2 concentration to obtain additional information on CO yields in all tests. However, analyzing the accumulated effluent to extract time dependent concentration data is likely to be accompanied by a high level of uncertainty. Thus, such calculations are not performed in this report.

The following is a repeat of the test numbering key, with format [X(Y)Zn]:

X: Fuel [S = sofas; B = bookcases; P = power cable]
Y: Q = heat release rate test (ceiling hole open)
Z: M = combustibles located near middle of burn room
    W = combustibles located near rear wall of burn room
    P = combustibles include PVC sheet
    C = doorway closed
    n = test number for that set of combustibles and location

A. Pre- and Post-flashover Time Intervals

There were four approaches from which to define a period of pre-flashover burning and one of post-flashover burning:

- From the test log. During the course of each test, the test manager noted the beginning and end of these two intervals.

  - The beginning of the pre-flashover period was designated when sustained burning of the fuel was observed followed the turning off and removal of the ignition burner.

  - The end of the pre-flashover period was designated as the test manager observed high intensity burning in the room and the approach of flashover approaching.

o   The beginning of the post-flashover period was flagged quickly after the point of flashover itself. This was most often identified by flaming of the crumpled newspaper telltale and by flames emerging from the doorway.

o   The end of the post-flashover period was recorded when the fuel began to be depleted and flames were no longer visible in the upper area of the doorway or by a decrease in the room upper layer temperature.

- From the event markers. These reflected the data system operator's response to the signals from the test manager and should produce similar time intervals.

- From the gas concentration data. During well ventilated flaming, the $[CO]/[CO_2]$ ratio should be 0.1 or lower and should be relatively steady. As flashover approaches, the ratio should increase and then plateau following flashover. As the test specimen burns out, the ratio should recede.

- From the temperature data. The time interval was narrowed when the thermocouples near the sampling probe indicated that the probe was not sampling from the upper layer.

The results of these considerations are shown in Table 11. Generally, we used the markers from the first approach if the middle three were similar (or didn't affect the results), then narrowed the interval, when warranted, by the temperature data. The smoke yield results were assumed to be insensitive to modest changes in the sampling interval.

B. Test Data

During the course of a series of experiments of this complexity, it is not uncommon for one or more instruments to malfunction during a given test. Table 12 lists the instrument failures and how we compensated for them (if at all).

In each of the following tables:

- *A blank cell indicates data that would have been derived from an uninstalled or non-functioning analyzer.*

- *A pound sign (#) indicates data that have been reconstructed as noted in Table 12.*

- *An asterisk (*) marks data derived from sampling that was not assuredly from the upper layer or for which the signal was too close to the background.*

Table 13 is a compilation of the calculated pre-flashover data:
- average mass loss rates from the test specimens,
- combustion product mass flows through the doorway, and
- volume fractions of major product gases at the four sampling locations.

Table 14 is a similar table of post-flashover data. It should be noted that tests SW1, SW2, and SW3 did not go to flashover. Their data in Table 14 is for later pre-flashover combustion.

Table 15 contains the calculated yields of the major gaseous combustion products from the open room tests.

Table 16 contains the calculated yields of the major gaseous combustion products from the closed room tests.

Table 17 contains the calculated soot yields.

Table 18 is a compilation of the calculated peak rates of heat release, combustion efficiencies, equivalence ratios, and concentration ratios of CO to CO_2.

Table 19 shows the level of consistency between the NDIR and FTIR analyzers.

Table 20 contains the results of calculations of the change in concentration of the combustion products along the length of the corridor.

Table 21 contains the calculated fractions of the notional yields of the product gases.

Table 22 shows the variance in yield values within sets of replicate tests.

The uncertainty in the measured concentrations varies with the magnitude of the concentration being measured and, for the FTIR measurements, the degree of spectral interference from other species. As noted in Section II.E.6, the uncertainties in the NDIR concentrations are generally low. The instrument linearities are well below 1 % of the measured value. The volume fraction noise level is about 1×10^{-5} for both the CO and CO_2 analyzers.

The FTIR uncertainties varied with the specific measurement and test. These are shown in Table 23.

Plots of the concentrations of the product gases at locations 2 and 4 are included in Appendix A.

Table 11. Time Intervals (s) over Which Analyses Were Performed

Test	From Test Log		From Data Collection		From Event Markers		From Gases		Used	
	Pre	Post	Pre	Post	Pre	Post	Pre	Post	Pre	Post
BW1	n.a. – n.a.	n.a. – n.a.	171 – 279	306 – 425	n.a. – n.a.	n.a. – n.a.	171 – 279	306 – 425	171 – 279	306 – 425
SW1	114 – 198	285 – 350	110 – 198	285 – 347	114 – 198	285 – 350	114 – 180	285 – 350	114 – 180	285 – 350
SW2	154 – 260	380 – 500	n.a. – 255	316 – 437	154 – 260	380 – 500	154 – 230	380 – 450	154 – 230	380 – 450
SW3	90 – 185	280 – 400	90 – 185	280 – 400	90 – 185	280 – 400	90 – 185	280 – 400	90 – 185	280 – 400
BW2	690 – 960	n.a. – n.a.	331 – 600	n.a. – n.a.	690 – 960	n.a. – n.a.	690 – 900	n.a. – n.a.	690 – 900	n.a – n.a
BW3	495 – 540	555 – 735	496 – n.a.	556 – 736	495 – 540	555 – 735	495 – 530	570 – 640	495 – 530	570 – 640
BW4	310 – 310	360 – 500	309 – 339	n.a. – n.a.	310 – 340	360 – 500	310 – 340	385 – 450	310 – 340	385 – 450
BW5	513 – 617	n.a. – n.a.	517 – 618	n.a. – n.a.	638 – 742	n.a. – n.a.	638 – 742	n.a – n.a	638 – 742	n.a – n.a
BW6	144 – 449	519 – 599	157 – n.a.	533 – n.a.	295 – 525	670 – 750	295 – 525	670 – 750	295 – 525	670 – 750
PQ1	149 – 214	306 – 486	148 – 214	306 – 485	59 – 125	217 – 396	59 – 115	250 – 396	59 – 115	250 – 396
PQ2	81 – 402	624 – 749	79 – 400	621 – 749	81 – 402	624 – 749	81 – 330	624 – 749	81 – 330	624 – 749
PW1	280 – 400	490 – 670	279 – 399	488 – 669	280 – 400	490 – 700	280 – 380	490 – 700	280 – 380	490 – 700
PW2	n.a. – 520	770 – 950	90 – 521	771 – 951	90 – 520	770 – 950	90 – 480	810 – 950	90 – 480	810 – 950
SW10	65 – 111	170 – 250	65 – 111	170 – 250	65 – 111	170 – 250	65 – 111	170 – 220	65 – 111	170 – 220
SW11	56 – 110	188 – 264	56 – 111	188 – 265	56 – 110	188 – 264	56 – 110	188 – 230	56 – 110	188 – 230
SW12	91 – 156	232 – 311	96 – 160	235 – 316	91 – 156	232 – 311	91 – 156	250 – 280	91 – 156	250 – 280
SW13	58 – 230	285 – 385	58 – 231	286 – 382	33 – 58	285 – 385	33 – 55	305 – 355	33 – 55	305 – 355
SW14	182 – 262	313 – 478	175 – 255	306 – 471	182 – 262	313 – 478	182 – 250	313 – 380	182 – 250	313 – 380
BP1	219 – 315	342 – 475	219 – 317	341 – 476	219 – 315	342 – 475	219 – 285	342 – 475	219 – 285	342 – 475
BP2	389 – 740	1453 – 1633	389 – 740	1455 – 1632	389 – 740	1453 – 1633	389 – 740	1453 – 1570	389 – 740	1453 – 1570
SC1	68 – 509	n.a. – n.a.	68 – 509	n.a. – n.a.	68 – 509	n.a. – n.a.	68 – 509	n.a. – n.a.	68 – 509	n.a – n.a
SC2	480 – n.a.	n.a. – n.a.	n.a. – n.a.	n.a. – n.a.	0 – 600	n.a. – n.a.	0 – 600	n.a. – n.a.	0 – 600	n.a – n.a
BW7	240 – 650	675 – 915	n.a. – n.a.	n.a. – n.a.	240 – 650	710 – 915	240 – 650	710 – 915	240 – 650	710 – 915
BP3	240 – 420	435 – 555	n.a. – n.a.	n.a. – n.a.	240 – 420	435 – 555	240 – 420	435 – 555	240 – 420	435 – 555

n.a.: not available

Table 12. Occurrence and Consequences of Malfunctioning Instruments

Test #	Installed Instruments	Malfunctioning Instrument(s)	Calculation Notes
SW1	All	None	Both collection periods provide pre-flashover calculations
SW2	All	None	Both collection periods provide pre-flashover calculations
SW3	All	None	Both collection periods provide pre-flashover calculations
BW2	Hood, Location 1 & 2 NDIR	None	
BW3	All but FTIR	None	
BW4	All	None	
BW5	Hood, load cell, Location 1 NDIR	None	
BW6	All but FTIR	None	
PQ1	All	None	
PQ2	All	None	
PW1	All	None	
PW2	All	Load cell	Adjusted PW2 loss rate such that the CO_2 yield equaled that from test PW1.
		Location 1 sampling, post-flashover	Discarded data.
SW10	All	None	
SW11	All	Velocity probes at 0.98 m and 0.83 m and doorway temperature 1.58 m from floor	Doorway flows calculated from operating velocity and temperature instruments in doorway
SW12	All	None	
SW13	All	Load cell "stuck" during pre-flashover period (low mass loss)	Mass loss set equal to the median from tests SW10, SW11, SW12, and SW14

53

Test #	Installed Instruments	Malfunctioning Instrument(s)	Calculation Notes
SW14	All	Load cell "stuck" during pre-flashover period (low mass loss)	Pre-flashover mass loss rate estimated from average mass loss of SW1-SW3, SW10-SW13
BP1	All	None	
BP2	All	PVC load cell	Substitute PVC mass loss curve from BP1 normalized by the ratio of the mass loss curves from the main load cell.
		Filter on CO/CO_2 sampling train at Location 2 partially clogged during post-flashover period	Post-flashover NDIR gas concentrations shifted by 30 s to correlate with downstream data
SC1	Location 1 analyzers; load cell	None	
SC2	Location 1 analyzers, load cell	None	
BW7	All	Location 1 sampling, post-flashover	Discarded data.
BP3	All	PVC load cell; Location 1 instruments	Substitute PVC mass loss curve from BP1 normalized by the ratio of the mass loss curves from the main load cell.

54

Table 13A. Pre-flashover Test Results: Mass Loss Rates and Doorway Flows

Test →	Mass loss rate (kg/s)		Doorway Flows (kg/s)						
	Main Fuel	PVC Sheet	Total	CO_2 NDIR	CO_2 FTIR	CO NDIR	CO FTIR	HCl FTIR	HCN FTIR
SW1	7.22E-03		6.51E-01	1.50E-03	1.59E-03	5.60E-05	2.15E-05	1.77E-05	2.55E-06
SW2	1.06E-02		6.73E-01	1.49E-03	1.07E-03	1.67E-05	1.17E-05	5.88E-06	8.66E-06
SW3	1.80E-02		6.88E-01	2.33E-03	4.16E-04	9.29E-06	1.18E-05	7.51E-06	5.62E-06
BW2	1.06E-02		7.77E-01	1.05E-03		4.95E-05			
BW3	8.83E-02		1.03E+00	4.73E-03		1.34E-04			
BW4	7.67E-02		8.32E-01	3.18E-02	1.89E-02	9.08E-04	3.18E-04	4.18E-05	3.29E-05
BW5	1.52E-03		4.85E-01	7.35E-04		1.94E-05			
BW6	3.20E-03		7.81E-01	1.85E-03		1.50E-04			
PQ1	2.60E-02		7.03E-01	2.05E-03	2.60E-04	1.21E-04	8.08E-06	1.29E-06	4.10E-06
PQ2	2.48E-03		3.40E-01	5.06E-04	4.73E-04	2.69E-05	1.56E-05	1.49E-05	1.42E-06
PW1	2.53E-02		7.11E-01	2.68E-03	2.07E-03	1.73E-04	8.04E-05	2.42E-04	7.82E-06
PW2	3.41E-02#		1.39E+00	3.62E-03	2.55E-03	1.00E-04	1.23E-04	1.48E-04	3.44E-05
SW10	1.48E-02		5.74E-01	1.04E-03	4.22E-04	9.66E-06	6.81E-06	7.18E-06	5.56E-06
SW11	1.10E-02		5.43E-01	3.19E-04	2.18E-04	3.54E-06	3.18E-07	2.91E-06	1.50E-06
SW12	6.94E-03		5.48E-01	1.43E-03	6.46E-04	7.34E-06	5.66E-06	7.51E-06	1.26E-06
SW13	1.51E-02#		3.76E-01	3.35E-04	8.68E-06	3.36E-06	5.76E-06	1.71E-06	5.14E-06
SW14	1.51E-02#		7.64E-01	1.49E-03	6.88E-04	6.81E-05	1.49E-05	1.03E-05	4.18E-06
BP1	4.76E-02	3.25E-03	8.64E-01	7.77E-03	2.39E-03	1.17E-04	4.52E-05	1.44E-18*	3.94E-06
BP2	1.23E-02	8.36E-04#	7.13E-01	1.38E-02	2.42E-02	5.37E-04	5.91E-04	8.61E-05	4.35E-05
SC1	4.08E-03								
SC2	2.92E-03								
BW7	1.14E-02		5.58E-01	6.14E-03	1.15E-02	2.61E-04	2.76E-04	4.27E-05	5.66E-06
BP3	2.87E-02	1.96E-03#	7.35E-01	7.15E-02	1.14E-02	1.91E-04	1.91E-04	2.84E-05	1.00E-05

Table 13B. Pre-flashover Test Results: Volume Fractions of Product Gases at Locations 1 and 2

Test	Location 1						Location 2					
	CO_2 NDIR	CO_2 FTIR	CO NDIR	CO FTIR	HCl FTIR	HCN FTIR	CO_2 NDIR	CO_2 FTIR	CO NDIR	CO FTIR	HCl FTIR	HCN FTIR
SW1	1.45E-02		3.24E-04				2.77E-03	2.76E-03	1.65E-04	5.98E-05	4.07E-05	7.73E-06
SW2	2.17E-02		3.45E-04				2.10E-03	1.48E-03	3.14E-05	2.28E-05	7.81E-06*	1.58E-05*
SW3	2.46E-02		5.13E-04				3.19E-03	5.52E-04	2.04E-05	2.52E-05	1.26E-05*	1.22E-05*
BW2	n.a.		n.a.				6.90E-04		5.18E-05			
BW3	1.31E-01		1.26E-02				6.26E-03		2.70E-04			
BW4	1.32E-01		1.12E-02				8.83E-03	5.25E-03	3.96E-04	1.39E-04	1.40E-05*	1.49E-05*
BW5	9.13E-03		6.22E-04				1.03E-03		4.21E-05			
BW6	7.96E-03		8.60E-04				1.25E-03		1.75E-04			
PQ1	2.96E-02		2.34E-03				3.42E-03	3.51E-04	3.21E-04	1.70E-05	2.68E-06*	1.17E-05*
PQ2	6.28E-03		8.05E-04				2.83E-03	2.66E-03	2.30E-04	1.33E-04	8.26E-05	1.27E-05*
PW1	2.61E-02		2.99E-03				4.51E-03	3.67E-03	4.53E-04	2.28E-04	5.01E-04	2.55E-05
PW2	4.73E-03		2.74E-04				1.86E-03	1.37E-03	8.24E-05	1.01E-04	9.28E-05	2.83E-05
SW10	3.19E-02		2.68E-04				1.83E-03	7.45E-04	2.65E-05	1.93E-05	1.62E-05*	1.45E-05*
SW11	2.28E-02		3.28E-04				1.87E-03	1.00E-03	3.64E-05	3.26E-06*	2.07E-05*	1.22E-05*
SW12	1.16E-02		2.22E-04				2.58E-03	1.17E-03	2.17E-05	1.62E-05	1.61E-05*	4.01E-06*
SW13	4.24E-03		7.05E-05				7.13E-04	2.14E-05	7.06E-06	2.05E-05	4.29E-06*	1.72E-05*
SW14	2.22E-02		2.03E-04				1.41E-03	6.66E-04	1.02E-04	2.31E-05	1.21E-05*	6.60E-06*
BP1	5.67E-02		1.47E-03				7.58E-03	2.25E-03	1.81E-04	6.54E-05		6.22E-06*
BP2	2.97E-02		2.19E-03				3.07E-03	3.95E-03	1.63E-04	1.47E-04	1.80E-05*	1.38E-05*
SC1	6.40E-02	4.13E-02	3.14E-03	9.82E-04	2.40E-05	1.87E-04						
SC2	5.38E-02	3.35E-02	2.52E-03	5.01E-04	5.03E-05	1.26E-04						
BW7							3.36E-03	5.87E-03	2.14E-04	2.18E-04	2.40E-05	5.09E-06*
BP3							9.06E-02	1.46E-02	3.78E-04	3.76E-04	4.30E-05	2.08E-05*

56

Table 13C. Pre-flashover Test Results: Volume Fractions of Product Gases at Locations 3 and 4

Test	Location 3		Location 4					
	CO_2 NDIR	CO NDIR	CO_2 NDIR	CO_2 FTIR	CO NDIR	CO FTIR	HCl FTIR	HCN FTIR
SW1	3.30E-03	1.95E-05	4.63E-03	2.76E-03	1.49E-04	5.98E-05	4.07E-05	7.73E-06*
SW2	4.51E-03	2.56E-05	7.64E-03	1.48E-03	5.88E-05	2.28E-05	7.81E-06*	1.58E-05*
SW3	4.61E-03	5.52E-05	8.65E-03	5.52E-04	5.87E-05	2.52E-05	1.26E-05*	1.22E-05*
BW2								
BW3	1.54E-02	3.60E-04	3.85E-02		3.28E-03			
BW4	1.49E-02	3.56E-04	3.58E-02	5.25E-03	2.56E-03	1.39E-04	1.40E-05*	1.49E-05*
BW5								
BW6	2.14E-03	1.09E-04	4.29E-03		3.79E-04			
PQ1	1.45E-03			3.51E-04		1.70E-05	2.68E-06*	1.17E-05*
PQ2	1.91E-03			2.66E-03		1.33E-04	8.26E-05	1.27E-05*
PW1	4.72E-03	4.62E-04	1.11E-02	3.67E-03	1.27E-03	2.28E-04	5.01E-04	2.55E-05
PW2	1.71E-03	8.16E-05	2.93E-03	1.37E-03	2.37E-04	1.01E-04	9.28E-05	2.83E-05
SW10	1.97E-03	7.24E-06*	1.21E-02	7.45E-04	1.87E-04	1.93E-05	1.62E-05*	1.45E-05*
SW11	1.73E-03	7.96E-06*	8.98E-03	1.00E-03	1.56E-04	3.26E-06	2.07E-05*	1.22E-05*
SW12	2.52E-03	9.58E-06*	6.89E-03	1.17E-03	1.62E-04	1.62E-05	1.61E-05*	4.01E-06*
SW13	7.44E-04		1.29E-03	2.14E-05	1.12E-04	2.05E-05	4.29E-06*	1.72E-05*
SW14	1.82E-03	2.58E-05	9.55E-03	6.66E-04	3.10E-04	2.31E-05	1.21E-05*	6.60E-06*
BP1	1.11E-02	2.29E-04	3.23E-02	2.25E-03	1.11E-03	6.54E-05		6.22E-06*
BP2	6.56E-03	4.00E-04	1.30E-02	3.95E-03	9.43E-04	1.47E-04	1.80E-05*	1.38E-05*
SC1								
BW7	1.71E-04		1.11E-02	5.87E-03	6.50E-04	2.18E-04	2.40E-05	5.09E-06*
BP3			1.67E-02	1.46E-02	8.28E-04	3.76E-04	4.30E-05	2.08E-05

Table 14A. Post-flashover Test Results: Mass Loss Rates and Doorway Flows

Test ↓	Mass loss rate (kg/s)			Doorway Flows (kg/s)					
	Main Fuel	PVC Sheet	Total	CO_2 NDIR	CO_2 FTIR	CO NDIR	CO FTIR	HCl FTIR	HCN FTIR
SW1	1.97E-02		8.33E-01	3.02E-02	3.50E-02	4.32E-04	3.41E-04	2.72E-04	2.15E-02
SW2	1.59E-02		8.33E-01	1.69E-02	2.36E-02	1.47E-04	1.48E-04	2.41E-04	1.96E-05
SW3	2.16E-02		8.72E-01	2.84E-02	4.44E-02	2.86E-04	3.34E-04	5.56E-04	1.07E-04
BW2									
BW3	1.16E-01		1.17E+00	1.02E-01		6.54E-03			
BW4	9.91E-02		1.11E+00	4.65E-01	1.29E-01	2.76E-02	5.22E-03	7.45E-05	3.60E-04
BW5									
BW6	1.99E-01		1.20E+00	1.13E-01		7.39E-03			
PQ1	8.31E-02		1.02E+00	1.08E-01	1.25E-01	1.23E-02	1.51E-02	1.92E-02	4.94E-04
PQ2	6.67E-02		8.69E-01	7.17E-02	7.72E-02	7.69E-03	8.60E-03	1.15E-02	2.16E-04
PW1	6.80E-02		9.75E-01	9.02E-02	1.10E-01	1.03E-02	1.05E-02	1.47E-02	2.13E-04
PW2	1.04E-01		1.67E+00	1.38E-01	1.76E-01	1.56E-02	1.62E-02	2.29E-02	4.00E-04
SW10	6.83E-02		1.06E+00	7.56E-02	7.22E-02	4.01E-03	3.93E-03	3.74E-04	1.27E-03
SW11	6.28E-02		1.04E+00	4.87E-02	4.92E-02	2.11E-03	1.66E-03	3.47E-04	5.25E-04
SW12	7.49E-02		1.11E+00	7.14E-02	9.09E-02	3.95E-03	4.15E-03	3.66E-04	1.29E-03
SW13	6.20E-02		1.09E+00	7.87E-02	7.09E-02	3.38E-03	2.82E-03	6.03E-04	1.03E-03
SW14	7.33E-02		1.26E+00	9.87E-02	1.21E-01	4.85E-03	4.33E-03	3.09E-04	1.20E-03
BP1	1.72E-01	5.06E-03	1.02E+00	1.11E-01	1.08E-01	7.94E-03	5.73E-03	8.47E-03	2.23E-04
BP2	9.98E-02	2.94E-03#	1.48E+00	2.14E-01	2.36E-01	8.08E-03	7.68E-03	1.43E-02	4.02E-04
SC1									
BW7	1.11E-01		1.63E+00	2.59E-01	1.69E-01	6.34E-03	2.79E-03	4.01E-04	1.51E-04
BP3	2.13E-01	6.27E-3#	1.42E+00	5.15E-02	3.99E-02	1.12E-03	1.28E-03	1.62E-03	8.64E-05

Tests SW1, SW2, and SW3 did not go to flashover. Their data are for later pre-flashover combustion.

58

Table 14B. Post-flashover Test Results: Volume Fractions of Product Gases at Locations 1 and 2

Test	Location 1						Location 2					
	CO_2 NDIR	CO_2 FTIR	CO NDIR	CO FTIR	HCl FTIR	HCN FTIR	CO_2 NDIR	CO_2 FTIR	CO NDIR	CO FTIR	HCl FTIR	HCN FTIR
BW1	n.a.		n.a.				1.31E-01		1.15E-02			
SW1	4.48E-02		8.56E-04				3.10E-02	3.56E-02	6.89E-04	5.45E-04	3.36E-04	1.39E-04
SW2	3.31E-02		5.92E-04				1.49E-02	2.39E-02	2.06E-04	2.42E-04	2.95E-04	3.22E-05
SW3	4.18E-02		7.88E-04				2.52E-02	3.96E-02	4.04E-04	4.70E-04	5.97E-04	1.59E-04
BW2												
BW3	1.79E-01		1.16E-02				1.20E-01		1.20E-02			
BW4	1.71E-01		1.39E-02				1.27E-01	3.52E-02	1.18E-02	2.23E-03	2.42E-05	1.58E-04
BW5												
BW6	1.66E-01		1.39E-02				9.74E-02		9.98E-03			
PQ1	1.28E-01		1.18E-02				7.43E-02	8.59E-02	1.33E-02	1.62E-02	1.59E-02	5.49E-04
PQ2	1.21E-01		9.45E-03				7.23E-02	7.78E-02	1.24E-02	1.37E-02	1.40E-02	3.57E-04
PW1	8.98E-02		1.11E-02				6.74E-02	8.19E-02	1.20E-02	1.22E-02	1.33E-02	2.62E-04
PW2	2.87E-05*		3.71E-05*				7.04E-02	8.99E-02	1.25E-02	1.32E-02	1.42E-02	3.38E-04
SW10	1.36E-01		5.66E-03				7.11E-02	6.82E-02	5.96E-03	5.84E-03	4.10E-04	1.96E-03
SW11	1.14E-01		4.04E-03				6.55E-02	6.63E-02	4.82E-03	3.95E-03	9.25E-04	1.53E-03
SW12	1.28E-01		4.90E-03				6.60E-02	8.43E-02	5.76E-03	6.07E-03	4.02E-04	1.96E-03
SW13	1.10E-01		1.82E-03				7.60E-02	6.85E-02	5.17E-03	4.36E-03	6.93E-04	1.65E-03
SW14	1.28E-01		3.87E-03				7.73E-02	9.51E-02	6.03E-03	5.36E-03	2.83E-04	1.55E-03
BP1	1.63E-01		1.42E-02				1.14E-01	1.12E-01	1.27E-02	9.17E-03	1.11E-02	3.67E-04
BP2	1.66E-01		2.37E-02				8.17E-02	8.73E-02	5.94E-03	5.49E-03	8.46E-03	2.91E-04
SC1												
BW7	3.07E-01						1.11E-01	7.25E-02	3.89E-03	1.71E-03	1.89E-04	9.59E-05
BP3							9.08E-02	8.91E-02	5.89E-03	4.69E-03	6.12E-03	3.35E-04

Tests SW1, SW2, and SW3 did not go to flashover. Their data are for later pre-flashover combustion.

59

Table 14C. Post-flashover Test Results: Volume Fractions of Product Gases at Locations 3 and 4

Test →	Location 3		Location 4					
	CO$_2$ NDIR	CO NDIR	CO$_2$ NDIR	CO$_2$ FTIR	CO NDIR	CO FTIR	HCl FTIR	HCN FTIR
BW1								
SW1	1.61E-02	2.09E-04	2.10E-02	4.16E-02	4.34E-04	2.20E-04	1.51E-05*	1.14E-05*
SW2	1.14E-02	1.19E-04	1.56E-02	3.62E-02	2.02E-04	1.28E-04	7.66E-05	3.78E-07*
SW3	1.58E-02	2.25E-04	1.94E-02	4.25E-02	2.42E-04	1.10E-04	1.47E-06*	1.18E-08*
BW2								
BW3	4.49E-02	5.00E-03	5.87E-02		5.95E-03			
BW4	4.29E-02	3.18E-03	5.60E-02	2.57E-02	4.35E-03	1.26E-03	2.51E-05	6.76E-05
BW5								
BW6	3.82E-02	1.45E-03	5.46E-02		4.97E-03			
PQ1	3.09E-02							
PQ2	2.94E-02							
PW1	3.02E-02	5.25E-03	3.82E-02	2.98E-02	6.98E-03	1.32E-03	2.62E-03	1.20E-04
PW2	3.20E-02	5.42E-03	4.18E-02	3.01E-02	6.87E-03	1.57E-03	3.03E-03	1.44E-04
SW10	3.89E-02	3.18E-03	4.71E-02	2.83E-02	2.79E-03	8.33E-04	2.05E-05*	5.72E-04
SW11	2.13E-03	3.83E-02	3.95E-02	3.77E-02	4.52E-03	1.99E-04	6.02E-05	1.78E-04
SW12	3.97E-02	3.01E-03	4.85E-02	2.58E-02	2.36E-03	2.06E-04	1.12E-05*	1.49E-04
SW13	3.86E-02	2.30E-03	4.52E-02	3.73E-02	1.84E-03	4.61E-04	1.21E-04	4.53E-04
SW14	3.91E-02	2.92E-03	4.92E-02	1.82E-02	3.10E-03	2.18E-04		1.40E-04
BP1	4.90E-02	5.59E-03	6.23E-02	2.62E-02	6.25E-03	9.60E-04	5.68E-04	8.11E-05
BP2	4.80E-02	3.08E-03	5.11E-02	2.70E-02	3.88E-03	6.61E-04	2.03E-03	4.17E-05
SC1								
SC2								
BW7	1.76E-04		4.78E-02	6.65E-04*	1.13E-03	9.32E-06*	6.08E-06*	2.15E-06*
BP3			5.77E-02	1.94E-02	6.55E-03	3.57E-04	2.13E-04	1.07E-05*

Tests SW1, SW2, and SW3 did not go to flashover. Their data are for later pre-flashover combustion.

60

Table 15. Yields of Combustion Products for Open Room Tests

Test	Pre-flashover						Post-flashover					
	CO NDIR	CO$_2$ NDIR	CO FTIR	CO$_2$ FTIR	HCl FTIR	HCN FTIR	CO NDIR	CO$_2$ NDIR	CO FTIR	CO$_2$ FTIR	HCl FTIR	HCN FTIR
SW1	2.20E-02	1.54E+00	1.73E-02	1.78E+00	1.38E-02	4.28E-03						
SW2	9.26E-03	1.06E+00	9.34E-03	1.49E+00	1.52E-02*	1.24E-03*						
SW3	1.32E-02	1.32E+00	1.55E-02	2.06E+00	2.57E-02*	4.95E-03*						
BW2	4.67E-03	9.89E-02										
BW3	1.51E-03	5.35E-02					5.63E-02	8.80E-01				
BW4	1.18E-02	4.14E-01	4.14E-03	2.46E-01	5.45E-04*	4.29E-04*	2.78E-01	4.69E+00	5.26E-02	1.30E+00	7.51E-04	3.63E-03
BW5	1.28E-02	4.83E-01										
BW6	4.71E-02	5.80E-01					3.72E-02	5.71E-01				
PQ1	4.67E-03	7.86E-02	3.10E-04	1.00E-02	4.97E-05*	1.58E-04*	1.48E-01	1.30E+00	1.82E-01	1.51E+00	2.31E-01	5.95E-03
PQ2	1.09E-02	2.04E-01	6.29E-03	1.91E-01	6.00E-03	5.73E-04*	1.15E-01	1.07E+00	1.29E-01	1.16E+00	1.72E-01	3.24E-03
PW1	6.82E-03	1.06E-01	3.18E-03	8.17E-02	9.57E-03	3.09E-04	1.51E-01	1.33E+00	1.55E-01	1.62E+00	2.16E-01	3.12E-03
PW2	2.94E-03#	1.06E-01#	3.60E-03#	7.49E-02#	4.34E-03#	1.01E-03#	1.50E-01	1.33E+00	1.56E-01	1.70E+00	2.20E-01	3.85E-03
SW10	6.54E-04	7.03E-02	4.61E-04	2.85E-02	4.86E-04*	3.76E-04*	5.87E-02	1.11E+00	5.75E-02	1.06E+00	5.47E-03	1.86E-02
SW11	3.13E-03	2.77E-01	4.61E-04	1.86E-01	2.95E-03*	1.55E-03*	3.37E-02	7.75E-01	2.64E-02	7.84E+01	5.53E-03	8.36E-03
SW12	1.06E-03	2.06E-01	8.16E-04	9.32E-02	1.08E-03*	1.81E-04*	5.28E-02	9.53E-01	5.54E-02	1.21E+00	4.89E-03	1.72E-02
SW13	2.24E-04#	2.23E-02#	3.84E-04#	5.78E-02#	1.14E-04#*	3.42E-04#*	5.45E-02	1.27E+00	4.55E-02	1.14E+00	9.73E-03	1.66E-02
SW14	4.55E-03#	9.97E-02#	9.98E-04#	4.60E-02#	6.89E-04#*	2.79E-04#*	6.62E-02	1.35E+00	5.92E-02	1.65E+00	4.22E-03	1.64E-02
BP1	2.31E-03	1.53E-01	8.89E-04	4.69E-02		7.76E-05*	4.49E-02	6.25E-01	3.17E-02	2.77E-01	1.67E+00	3.63E-01
BP2	4.11E-02	1.05E+00	4.51E-02	1.85E+00	6.58E-03#*	3.33E-03*	7.87E-02	2.09E+00	3.00E-02	4.67E-01	4.87E+00#	5.12E-01
SC1												
SC2												
BW7	2.30E-02	5.41E-01	2.43E-02	1.01E+00	3.76E-03	4.99E-04*	5.74E-02	2.35E+00	2.52E-02	1.53E+00	3.63E-03	1.36E-03
BP3	6.65E-03	2.49E+00	6.66E-03	3.98E-01	9.90E-04#	3.49E-04*	5.26E-03	2.42E-01	6.04E-03	1.87E-01	2.58E-01#	4.06E-04

Note: Tests SW1, SW2, SW3 did not go to flashover; the later pre-flashover concentration data are higher than the early pre-flashover data, and the former have been inserted in the "Pre-flashover" half of this Table.

61

Table 16. Yields of Combustion Products from Closed Room Tests

SC1

Time (s)	Volume Fraction NDIR CO$_2$	Volume Fraction NDIR CO	Volume Fraction FTIR CO$_2$	Volume Fraction FTIR CO	Volume Fraction FTIR HCl	Volume Fraction FTIR HCN	Mass Lost (kg)	Upper Layer Depth (m)	Mean Upper Layer T (K)	Yield NDIR CO$_2$	Yield NDIR CO	Yield FTIR CO$_2$	Yield FTIR CO	Yield FTIR HCl	Yield FTIR HCN
150	3.60E-02	5.10E-04	2.40E-02	2.02E-04	7.50E-07	2.60E-06	0.44	1.85	498	1.45E+00	1.31E-02	9.65E-01	5.17E-03	2.43E-05	6.42E-05
200	6.13E-02	1.20E-03	3.90E-02	3.74E-04	2.20E-06	7.70E-05	0.84	1.85	493	1.31E+00	1.64E-02	8.36E-01	5.10E-03	3.80E-05	1.01E-03
250	7.80E-02	2.53E-03	5.04E-02	7.00E-04	1.54E-06	1.91E-04	1.17	1.85	488	1.21E+00	2.50E-02	7.83E-01	6.92E-03	1.93E-05	1.82E-03
300	8.61E-02	4.15E-03	5.95E-02	1.19E-03	2.67E-05	2.79E-04	1.40	1.85	473	1.15E+00	3.54E-02	7.97E-01	1.01E-02	2.89E-04	2.29E-03
350	8.75E-02	5.66E-03	4.60E-02	1.82E-03	4.50E-05	3.37E-04	1.60	1.85	443	1.10E+00	4.51E-02	5.76E-01	1.45E-02	4.54E-04	2.59E-03
400	8.14E-02	5.40E-03	4.10E-02	1.64E-03	3.90E-05	3.00E-04	1.70	1.85	435	9.77E-01	4.12E-02	4.92E-01	1.25E-02	3.78E-04	2.21E-03
450	7.68E-02	5.20E-03	5.85E-02	1.62E-03	4.82E-05	2.89E-04	1.76	1.85	408	9.49E-01	4.09E-02	7.23E-01	1.27E-02	4.81E-04	2.19E-03
500	7.09E-02	4.80E-03	5.18E-02	1.57E-03	4.35E-05	2.60E-04	1.83	1.85	403	8.53E-01	3.67E-02	6.23E-01	1.20E-02	4.22E-04	1.92E-03

SC2

Time (s)	Volume Fraction NDIR CO$_2$	Volume Fraction NDIR CO	Volume Fraction FTIR CO$_2$	Volume Fraction FTIR CO	Volume Fraction FTIR HCl	Volume Fraction FTIR HCN	Mass Lost (kg)	Upper Layer Depth (m)	Mean Upper Layer T (K)	Yield NDIR CO$_2$	Yield NDIR CO	Yield FTIR CO$_2$	Yield FTIR CO	Yield FTIR HCl	Yield FTIR HCN
150	2.90E-02	4.20E-04	1.06E-02	7.60E-05	1.33E-05	3.34E-07	0.24	1.75	463	2.19E+00	2.02E-02	8.01E-01	3.65E-03	8.10E-04	1.55E-05
200	5.33E-02	9.90E-04	2.17E-02	1.75E-04	8.50E-05	7.50E-06	0.62	1.85	488	1.56E+00	1.85E-02	6.36E-01	3.27E-03	2.01E-03	1.35E-04
250	7.06E-02	2.20E-03	3.05E-02	2.55E-04	1.60E-04	4.36E-05	0.92	1.85	473	1.44E+00	2.85E-02	6.22E-01	3.31E-03	2.63E-03	5.46E-04
300	7.64E-02	3.50E-03	4.70E-02	5.60E-04	1.04E-04	1.56E-04	1.11	1.85	448	1.36E+00	3.97E-02	8.39E-01	6.36E-03	1.50E-03	1.71E-03
350	7.71E-02	3.90E-03	5.40E-02	7.20E-04	3.70E-05	2.14E-04	1.27	1.85	438	1.23E+00	3.96E-02	8.61E-01	7.31E-03	4.76E-04	2.09E-03
400	7.56E-02	4.05E-03	4.80E-02	9.90E-04	4.20E-05	2.32E-04	1.41	1.85	423	1.12E+00	3.83E-02	7.14E-01	9.37E-03	5.04E-04	2.12E-03
450	7.26E-02	4.14E-03	5.67E-02	8.50E-04	4.40E-05	2.35E-04	1.49	1.85	418	1.03E+00	3.75E-02	8.08E-01	7.71E-03	5.06E-04	2.05E-03
500	7.34E-02	4.32E-03	5.36E-02	9.70E-04	4.60E-05	2.41E-04	1.61	1.85	413	9.79E-01	3.67E-02	7.15E-01	8.24E-03	4.95E-04	1.97E-03
550	7.18E-02	4.35E-03	5.72E-02	9.20E-04	4.80E-05	2.50E-04	1.70	1.85	408	9.18E-01	3.54E-02	7.32E-01	7.49E-03	4.95E-04	1.96E-03
600	6.95E-02	4.35E-03	5.78E-02	1.03E-03	5.10E-05	2.53E-04	1.76	1.85	395	8.87E-01	3.53E-02	7.38E-01	8.37E-03	5.25E-04	1.98E-03

62

Table 17. Smoke Yields

	Pre-flashover			Post-flashover			Sofas		Bookcases		Cable		Bookcase/PVC	
	Loc 4	Loc 2	4/2	Loc 4	Loc 2	4/2	A	B	A	B	A	B	A	B
SW1	2.5E-02	5.7E-02	4.4E-01	1.3E-02					1.9E+00					
SW2	2.5E-02	4.3E-01	5.9E-02	1.8E-02	1.2E-02	1.5E+00								
SW3				8.1E-03	8.7E-03	9.2E-01								
BW2														
BW3	1.1E-02	5.0E-02	2.2E-01	1.2E-02	7.0E-03	1.6E+00			9.6E-01	7.2E+00				
BW4	2.2E-02			9.9E-03	6.3E-03	1.6E+00			2.3E+00					
BW5														
BW6														
PQ1		1.5E-01			4.8E-02									
PQ2					7.0E-02									
PW1	8.8E-02	2.0E-01	4.5E-01	3.7E-02	8.3E-02	4.4E-01					2.4E+00	2.3E+00		
PW2	7.2E-02	2.7E-01	2.7E-01	4.2E-02	1.3E-01	3.1E-01					1.7E+00	2.0E+00		
SW10	9.7E-02	3.9E-01	2.5E-01		5.2E-02			7.6E+00						
SW11	4.5E-02	2.4E-01	1.9E-01		2.7E-02			8.8E+00						
SW12	2.9E-01	2.8E-01	1.1E+00	3.9E-02	5.0E-02	7.7E-01	7.6E+00	5.5E+00						
SW13	8.8E-02	2.2E-01	3.9E-01	3.1E-02	2.9E-02	1.1E+00	2.8E+00	7.8E+00						
SW14	7.7E-02			5.4E-02	6.8E-02	7.9E-01								
BP1	1.1E-02	5.9E-02	1.8E-01	6.5E-02	1.0E-01	6.2E-01							1.6E-01	5.7E-01
BP2	3.3E-02	5.9E-02	5.7E-01	9.0E-02	1.5E-01	6.2E-01							3.7E-01	4.0E-01
SC1														
SC2														
BW7	1.0E-02	2.9E-02	3.6E-01	8.5E-03	1.6E-02	5.2E-01			1.2E+00	1.7E+00	1.2E+00			
BP3	7.7E-03			1.2E-01										
Mean							5.2E+00	7.4E+00	1.6E+00	4.5E+00	1.8E+00	2.2E+00	2.7E-01	4.8E-01
σ							3.4E+00	1.4E+00	6.9E-01	3.8E+00	5.8E-01	2.2E-01	1.5E-01	1.2E-01

A: ratio of pre-flashover to post-flashover yield at position 4

B: same ratio at position 2

63

Table 18. Combustion Characterization

Test →	Peak Rate of Heat Release (MJ/kg)	Combustion Efficiency		Global Equivalence Ratio		[CO]/[CO₂] (location 2)	
		Pre-flashover	Post-flashover	Pre-flashover	Post-flashover	Pre-flashover	Post-flashover
BQW1	1880						
BQW2	190						
SQW3	1170						
SQW4	440						
SQM1	380						
SQM2	2010						
SW1		0.94		0.11		0.022	
SW2		0.98		0.09		0.014	
SW3		0.99		0.12		0.016	
BW2		0.93		0.04		0.084	
BW3		0.96	0.91	0.23	0.27	0.081	0.059
BW4		0.96	0.91	0.25	0.24	0.045	0.072
BW5		0.96		0.01		0.041	
BW6		0.89	0.91	0.01	0.44	0.139	0.103
PQ1	1510	0.91	0.85	0.16	0.35	0.093	0.176
PQ2	1390	0.92	0.86	0.03	0.33	0.083	0.171
PW1		0.91	0.85	0.15	0.30	0.103	0.179
PW2		0.96	0.85			0.039	0.178
SW10		0.99	0.92	0.12	0.31	0.014	0.064
SW11		0.98	0.93	0.10	0.19	0.019	0.053
SW12		0.99	0.92	0.06	0.32	0.007	0.060
SW13		0.98	0.94	0.11	0.27	0.014	0.053
SW14		0.93	0.93	0.00	0.28	0.076	0.045
BP1		0.98	0.90	0.18	0.54	0.035	0.111
BP2		0.94	0.94	0.05	0.21	0.053	0.065
SC1							
SC2							
BW7		0.94	0.96	0.05	0.18	0.064	0.035
BP3		1.00	0.97	0.12	0.46	0.004	0.065

Table 19. NDIR/FTIR Volume Fraction Ratios (Location 2)

Test ↓	Pre-flashover		Post-flashover	
	CO	CO_2	CO	CO_2
SW1	1.3	0.9		
SW2	1.0	0.7		
SW3	0.8	0.7		
BW2				
BW3				
BW4	3.0	1.7	5.3	3.6
BW5				
BW6				
PQ1	18.9	9.8	0.8	0.9
PQ2	1.7	1.1	1.0	0.9
PW1	1.9	1.2	1.0	0.8
PW2	1.3	6.2	1.1	0.8
SW10	1.4	2.8	1.0	1.0
SW11	12.3	2.2	1.6	1.0
SW12	1.4	2.2	1.0	0.8
SW13	0.4	33.6	1.2	1.1
SW14	4.5	2.1	1.2	0.8
BP1	6.8	8.2	1.9	1.1
BP2	16.1	5.7	1.0	0.9
SC1				
SC2				
BW7		1.1		1.5
BP3	1.1	7.5	1.2	1.5
Mean	4.7	4.9	1.15*	1.00*
σ	5.5	7.5	0.27	0.23

* Without BW4

Table 20. Ratios of Concentrations: Position 4 (Downstream)/Position 2 (Upstream)

	Pre-flashover							Post-flashover						
	CO	CO_2	CO	CO_2	HCl	HCN	Smoke	CO	CO_2	CO	CO_2	HCl	HCN	Smoke
	NDIR	NDIR	FTIR	FTIR	FTIR	FTIR		NDIR	NDIR	FTIR	FTIR	FTIR	FTIR	
SW1	0.63	0.68	0.55	1.16	*	*	0.71							
SW2	0.90	0.94	0.55	1.52	*	*	1.39							
SW3	0.67	0.78	0.25	1.14	*	*	0.70							
BW2	*	*	*	*	*	*								
BW3	12.12	6.15	*	*	*	*	0.76	0.49	0.49					
BW4	6.45	4.05	8.36	3.29	*	*	0.68	0.37	0.44	0.57	0.73	1.04	0.43	0.67
BW5			*	*	*	*								
BW6	2.17	3.42	*	*	*	*		0.50	0.56					
PQ1														
PQ2														
PW1	2.79	2.46	1.01	2.06	1.40	*	0.22	0.58	0.57	0.11	0.36	0.20	0.46	0.22
PW2	2.88	1.57	0.90	2.15	1.90	*	0.14	0.55	0.59	0.12	0.33	0.21	0.43	0.14
SW10	7.08	6.62	*	2.53	*	*	0.32	0.47	0.66	0.14	0.42	0.05	0.29	0.32
SW11	4.28	4.81	*	4.14	*	*	0.51	0.94	0.60	0.05	0.57	0.07	0.12	0.51
SW12	7.44	2.67	*	1.63	*	*	0.47	0.41	0.74	0.03	0.31	0.03	0.08	0.47
SW13	*	1.80	*	22.22	*	*	0.64	0.36	0.59	0.11	0.54	0.17	0.27	0.64
SW14	3.02	6.76	*	2.26	*	*	0.43	0.51	0.64	0.04	0.19		0.09	0.43
BP1	6.10	4.26	0.34	1.80	*	*	0.29	0.49	0.55	0.10	0.23	0.05	0.22	0.29
BP2	5.79	4.23	1.53	2.85	*	*	0.23	0.65	0.63	0.12	0.31	0.24	0.14	0.23
SC1														
SC2														
BW7	3.03	3.30	*	0.03	*	*	0.23	0.29	0.43	*	0.01	*	*	0.23
BP3	2.19	0.18	0.37	0.91	*	*	0.64	1.11	0.63	0.08	0.22	0.03	0.03	0.64

* Yield of the gas was sufficiently low that the ratio is meaningless.

66

Table 21. Fractions of Notional Yields

		CO_2	CO	HCl	HCN
Sofa	Post	0.57 ± 0.12	$(4.0 \pm 0.9) \times 10^{-2}$	0.86 ± 0.27	$(7.8 \pm 1.8) \times 10^{-2}$
	Pre	0.80 ± 0.17	$(1.13 \pm 0.35) \times 10^{-2}$	2.6 ± 0.8	$(1.81 \pm 0.83) \times 10^{-3}$
	Closed (200 s)[1]	0.72 ± 0.07	$(1.37 \pm 0.08) \times 10^{-2}$	$(7 \pm 1) \times 10^{-2}$	$(1.1 \pm 0.1) \times 10^{-1}$
Bookcase	Post	1.10 ± 0.80	$(4.2 \pm 1.2) \times 10^{-2}$	0.85 ± 0.55	$(4.4 \pm 2.0) \times 10^{-2}$
	Pre	0.29 ± 0.14	$(2.2 \pm 1.2) \times 10^{-2}$	0.85 ± 0.64	$(8.1 \pm 0.6) \times 10^{-3}$
Cable	Post	0.65 ± 0.10	0.111 ± 0.013	0.58 ± 0.06	0.100 ± 0.028
	Pre	$(5.7 \pm 2.4) \times 10^{-2}$	$(4.1 \pm 1.9) \times 10^{-3}$	$(1.78 \pm 0.48) \times 10^{-2}$	$(1.58 \pm 0.72) \times 10^{-2}$
PVC sheet	Post	[2]	[2]	4.3 ± 3.6	[3]
	Pre	[2]	[2]	[4]	[3]

[1] Earliest time with significant combustible mass loss and before significant vitiation. Calculations are for comparison with open-door pre-flashover results. See Section V.E for discussion of time-dependent results.

[2] Pre-flashover mass loss from the PVC sheet was negligible. Thus yields are not calculable.

[3] All HCN generated in the PVC sheet tests is from the bookcase and the results are included there.

[4] Mass loss and HCl concentrations were too low to obtain a meaningful value.

Table 22a. Variance in Product Yields Among Replicate Tests (SW 10 to SW14)

		CO_2	CO	HCl	HCN
Pre-flashover	NDIR	0.135 ± 0.092 (69 %)	$(1.92 \pm 1.65) \times 10^{-3}$ (86 %)	--	--
	FTIR	$(8.2 \pm 5.6) \times 10^{-2}$ (68 %)	$(6.2 \pm 2.4) \times 10^{-4}$ (39 %)	$(1.06 \pm 0.99) \times 10^{-3}$ (94 %)	$(5.5 \pm 5.1) \times 10^{-4}$ (92 %)
Post-flashover	NDIR	1.09 ± 0.21 (19 %)	$(5.3 \pm 1.1) \times 10^{-2}$ (21%)	--	--
	FTIR	1.17 ± 0.28 (24 %)	$(4.9 \pm 1.2) \times 10^{-2}$ (25 %)	$(6.0 \pm 1.9) \times 10^{-3}$ (32 %)	$(1.5 \pm 0.36) \times 10^{-2}$ (24 %)

Table 22b. Variance in Product Yields Among Replicate Tests (SW 1 to SW3)

		CO_2	CO	HCl	HCN
Pre-flashover	NDIR	1.31 ± 0.20 (15 %)	$(1.48 \pm 0.53) \times 10^{-2}$ (36 %)	--	--
	FTIR	1.78 ± 0.23 (13 %)	$(1.40 \pm 0.34) \times 10^{-2}$ (24 %)	$(1.82 \pm 0.53) \times 10^{-2}$ (29 %)	$(3.5 \pm 1.6) \times 10^{-3}$ (45 %)

Table 23a. Uncertainties in FTIR Volume Fractions of Product Gases; Pre-flashover Data. (Expressed as a percentage of the volume fractions reported in Table 12)

Test ↓	Location 1				Location 2				Location 4			
	CO_2	CO	HCl	HCN	CO_2	CO	HCl	HCN	CO_2	CO	HCl	HCN
SW1					± 0.5	± 5	± 12	± 43	± 0.2	± 7	± 4	± 75
SW2					± 0.1	± 4	± 4	± 8	± 0.2	± 3	± 200	± 60
SW3					± 5	± 3	± 17	± 11	± 0.4	± 15	± 29	± 64
BW2												
BW3												
BW4												
BW5												
BW6												
PQ1					± 0.5	± 5	± 31	± 9				
PQ2					± 0.7	± 0.7	± 1	± 7				
PW1					± 0.1	± 0.8	± 0.7	± 4	± 0.2	± 0.9	± 0.9	± 27
PW2					± 0.1	± 0.6	± 1	± 3	± 0.1	± 0.7	± 0.5	± 0.9
SW10					± 3	± 8	± 20	± 15	± 0.3	± 26	± 17	± 26
SW11					± 0.3	± 76	± 12	± 17	± 3	± 24	± 85	± 194
SW12					± 0.1	± 8	± 16	± 32	± 3	± 218	± 44	± 51
SW13					± 10	± 12	± 125	± 21	± 0.8	± 151	± 28	± 378
SW14					± 0.3	± 6	± 23	± 23	± 0.4	± 16	± 75	± 213
BP1					± 0.1	± 1	± 147	± 19	± 0.2	± 2	± 65	± 28
BP2					± 0.1	± 0.5	± 3	± 3	± 0.1	± 0.7	± 93	± 5
SC1	± 0.3	± 0.1	± 4	± 19	± 1	± 35	± 400	± 400				
SC2	± 0.5	± 0.1	± 4	± 4	± 1	± 35	± 500	± 62				
BW7					± 0.1	± 0.2	± 3	± 13	± 0.9	± 17	± 34	± 9
BP3					± 0.1	± 0.3	± 3	± 6	± 0.1	± 0.8	± 26	± 130

Table 23b. Uncertainties in FTIR Volume Fractions of Product Gases; Post-flashover Data. (Expressed as a percentage of the volume fractions reported in Table 13)

Test ↓	Location 2				Location 4			
	CO$_2$	CO	HCl	HCN	CO$_2$	CO	HCl	HCN
SW1	± 0.5	± 3	± 4	± 25				
SW2	± 0.2	± 0.6	± 2	± 8	± 0.2	± 3	± 44	± 44
SW3	± 0.1	± 0.3	± 0.5	± 2	± 0.2	± 4	± 10	± 209
BW2					± 0.2	± 4	± 162	± 162
BW3								
BW4								
BW5					± 0.2	± 0.2	± 25	± 9
BW6								
PQ2	± 0.1	± 0.1	± 0.2	± 6				
PW1	± 0.1	± 0.7	± 0.2	± 6				
PW2	± 0.1	± 0.1	± 0.2	± 7	± 0.1	± 0.6	± 0.9	± 12
SW10	± 0.2	± 0.1	± 0.6	± 2	± 0.1	± 1	± 1	± 7
SW11	± 0.2	± 0.2	± 4	± 3	± 0.25	± 25	± 25	± 4
SW12	± 0.2	± 0.2	± 5	± 2	± 0.3	± 28	± 28	± 32
SW13	± 0.2	± 0.2	± 4	± 2	± 0.3	± 316	± 32	± 19
SW14	± 0.2	± 0.2	± 2	± 0.8	± 0.3	± 100	± 21	± 6
BP1	± 0.1	± 0.1	± 0.1	± 3	± 0.2	± 17	± 112	± 14
BP2	± 0.1	± 0.9	± 1	± 11	± 0.2	± 2	± 2	± 30
SC1					± 0.2	± 1	± 2	± 112
SC2								
BW7	± 0.	± 0.4	± 1	± 9				
BP3	± 0.1	± 0.3	± 0.8	± 11	± 0.8	± 24	± 24	± 24

C. Checks on Data Reliability

There are certain ratios whose values are fixed or can be estimated. Examining these provides a first assessment as to the integrity of the data set

1. Cross-instrument Similarity

The ratio of the volume fraction obtained using NDIR to that measured using FTIR should be unity for either CO_2 or CO measured at the same location (data from Table 19). The data set for these ratios includes experiments involving bookcases, sofas and cables. The presented uncertainty is the standard deviation. The post-flashover values are the highest concentrations with the highest signal to noise and are thus the most likely to manifest proper behavior.

Much of the pre-flashover data from location 4 are too near background to assess agreement, and the vertical temperature profiles indicated that during the pre-flashover period, the location 4 sampling probes were not safely in the hot upper layer and thus were not assuredly sampling room fire effluent. Tests SW1, SW2 and SW2 did not reach flashover, but did continue to produce pre-flashover smoke that enveloped the location 2 sampling probes. Thus, the late data from these three tests constitute a fair check on cross-instrument similarity.

- For post-flashover measurements at location 2:
    - $[CO]_{NDIR}/[CO]_{FTIR} = 1.17 \pm 0.35$
    - $[CO_2]_{NDIR}/[CO_2]_{FTIR} = 0.97 \pm 0.19$
- For late pre-flashover measurements at location 2:
    - $[CO]_{NDIR}/[CO]_{FTIR} = 0.99 \pm 0.19$
    - $[CO_2]_{NDIR}/[CO_2]_{FTIR} = 0.71 \pm 0.11$
- For post-flashover measurements at location 4
    - $[CO]_{NDIR}/[CO]_{FTIR} = 5.7 \pm 3.7$
    - $[CO_2]_{NDIR}/[CO_2]_{FTIR} = 1.87 \pm 0.60$
- For late pre-flashover measurements at location 4
    - $[CO]_{NDIR}/[CO]_{FTIR} = 1.91 \pm 0.26$
    - $[CO_2]_{NDIR}/[CO_2]_{FTIR} = 0.46 \pm 0.03$

2. Location Similarity

The ratio of the volume fractions of CO_2 measured at two locations (Table 20) should reflect dilution only and thus should be the same for all tests and instruments. [See Section IV.E for the ratios for other gases.]

For post-flashover NDIR measurements:

- $[CO2]_{loc4}/[CO2]_{loc2}$ = 0.59 ± 0.07

For late pre-flashover NDIR measurements:

- $[CO2]_{loc4}/[CO2]_{loc2}$ = 0.83 ± 0.16

For post-flashover FTIR measurements:

- $[CO2]_{loc4}/[CO2]_{loc2}$ = 0.38 ± 0.16

For late pre-flashover FTIR measurements:

- $[CO2]_{loc4}/[CO2]_{loc2}$ = 1.25 ± 0.19

3. Notional Yield Fractions

From the post-flashover gas concentration data, most of the fuel carbon appears as CO_2, with lesser amounts appearing as CO and carbonaceous soot. Thus, one should expect the fraction of the notional yield of CO_2 calculated from the concentration measurements to be slightly less than unity. For Cl, a value less than unity probably represents losses to the walls combined with losses in the sampling line. The values for CO and HCN should be well under unity, as most of the C and N is expected to be found in other combustion products. Table 21 presents the results, with the FTIR and NDIR values combined in the calculations for CO_2 and CO.

D. Test Repeatability

It is well known that there are numerous sources of variability in real-scale fire tests. These could also impact the repeatability of the measured toxicant yields. Time and resources did not provide for an exhaustive evaluation of test repeatability. Table 22a shows the mean yields of the principal toxicants and the uncertainty (standard deviation) from five replicate tests (SW10-SW14) of one fuel and configuration. Figure 15 shows the mass loss curves for the five tests with the time scales shifted to a common ignition time.

Because the probes at location 2 were not always in the effluent stream early in a test, the later pre-flashover data for tests SW1-SW3 are also shown (Table 22b). Even though the number of cushions differed among these tests, the burning behavior was similar, as shown in Figure 16.

Figures 17 and 18 show the time-integrated yields of the measured gases for the two closed room tests. Figure 19 shows the evolving oxygen concentration (30 cm from the ceiling) for the two tests.

Figure 15. Mass Loss vs. Time: Tests SW10 through SW14

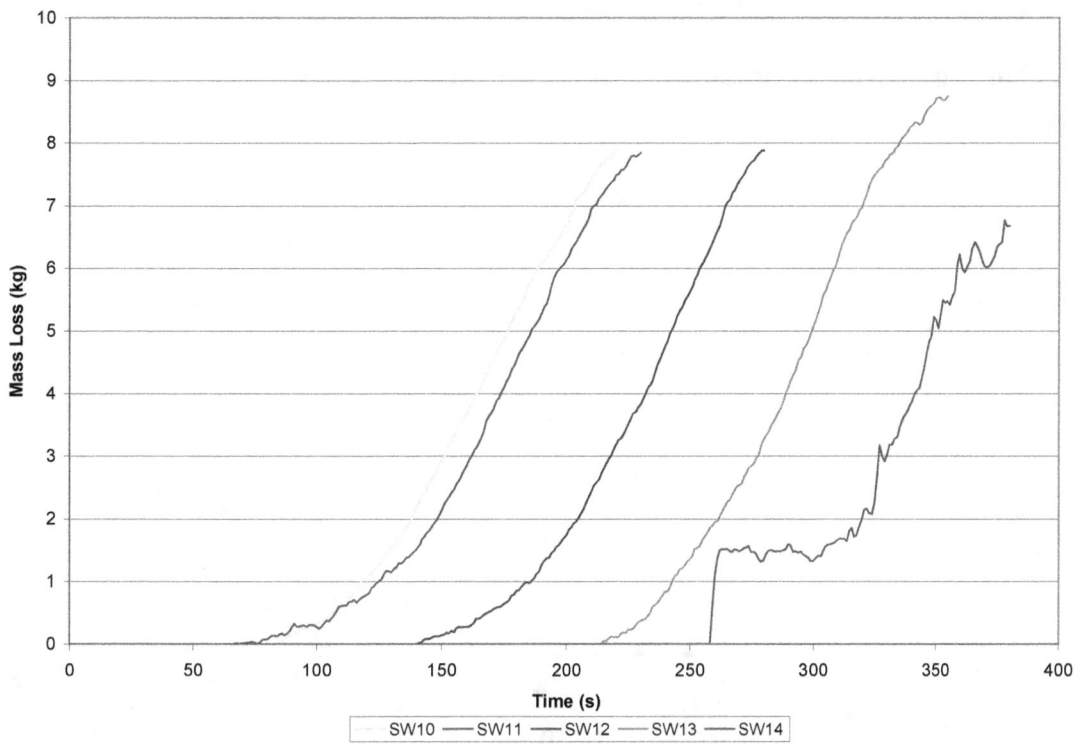

Figure 16. Mass Loss vs. Time: Tests SW1 through SW3

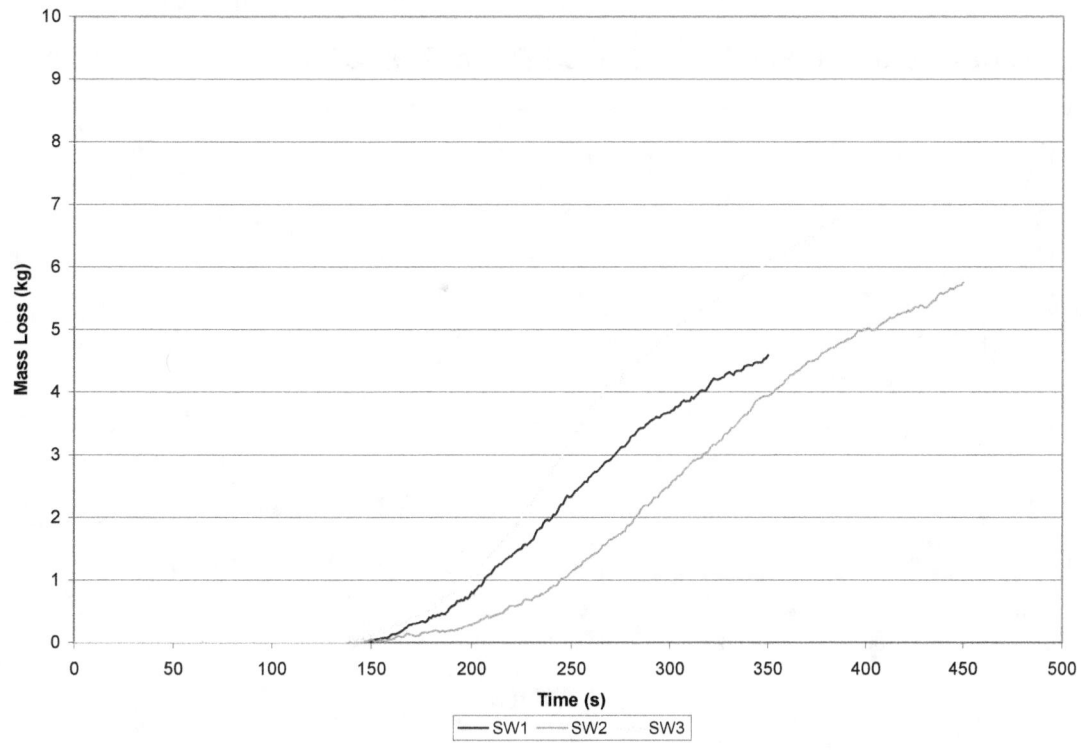

Figure 17. Time-integrated Yields for Closed-Room Test SC1

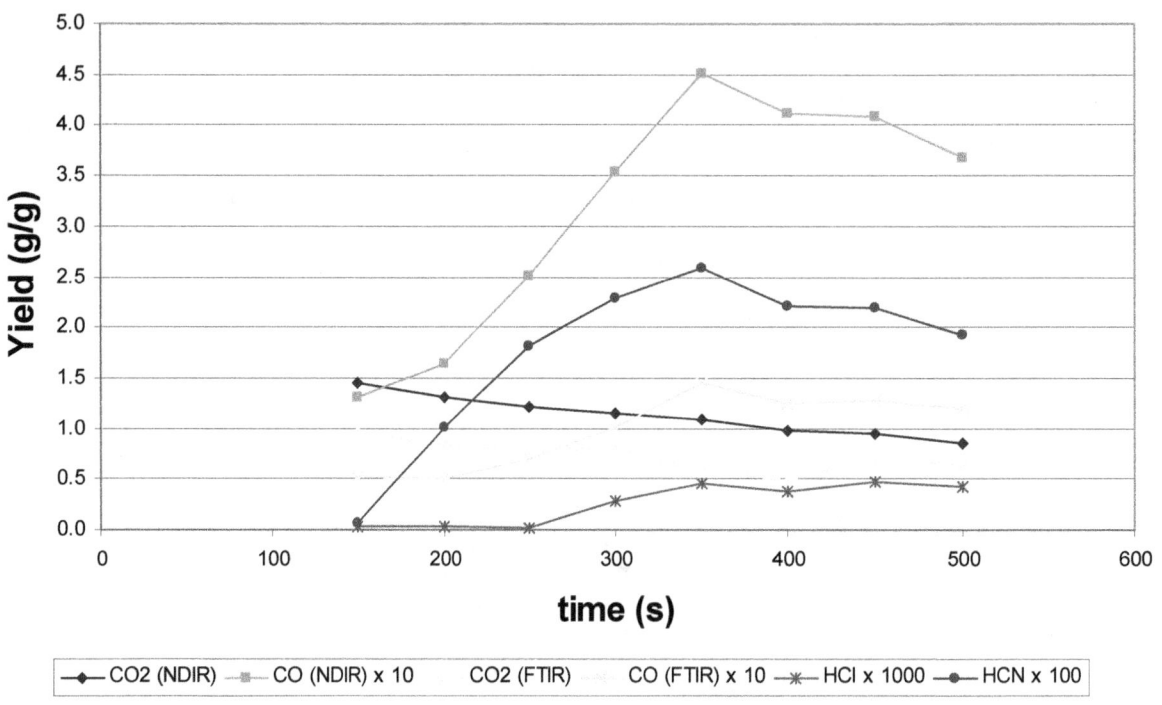

Figure 18. Time-integrated Yields for Closed-Room Test SC2

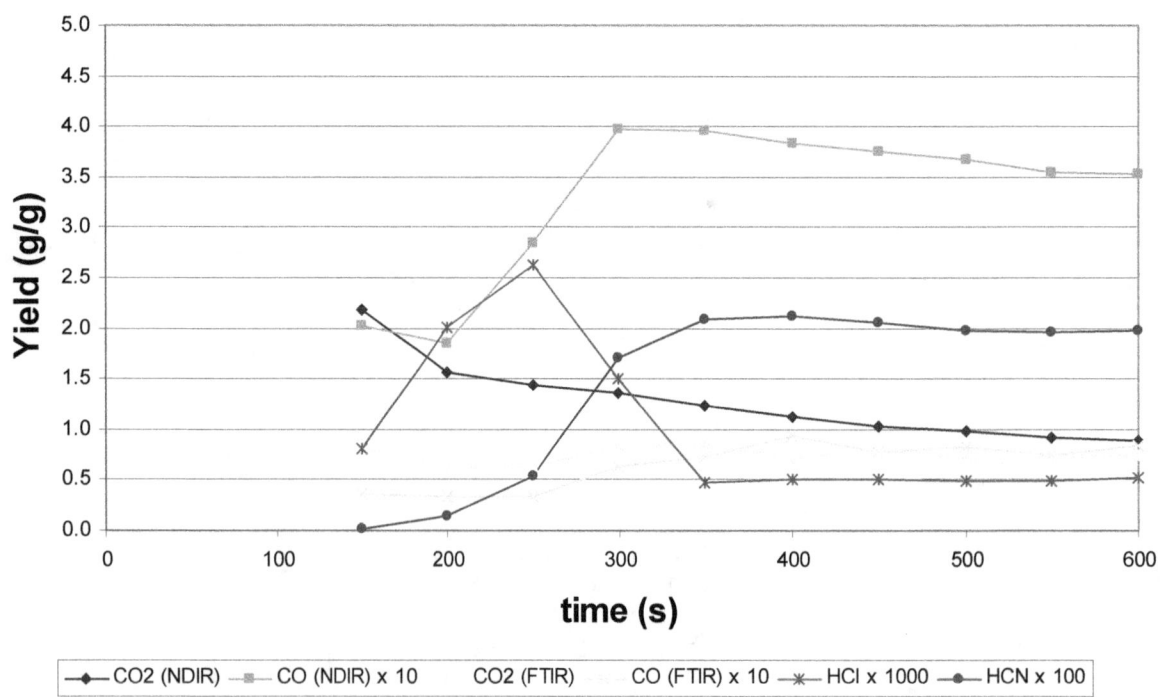

Figure 19. Oxygen Concentration in Closed Room Tests

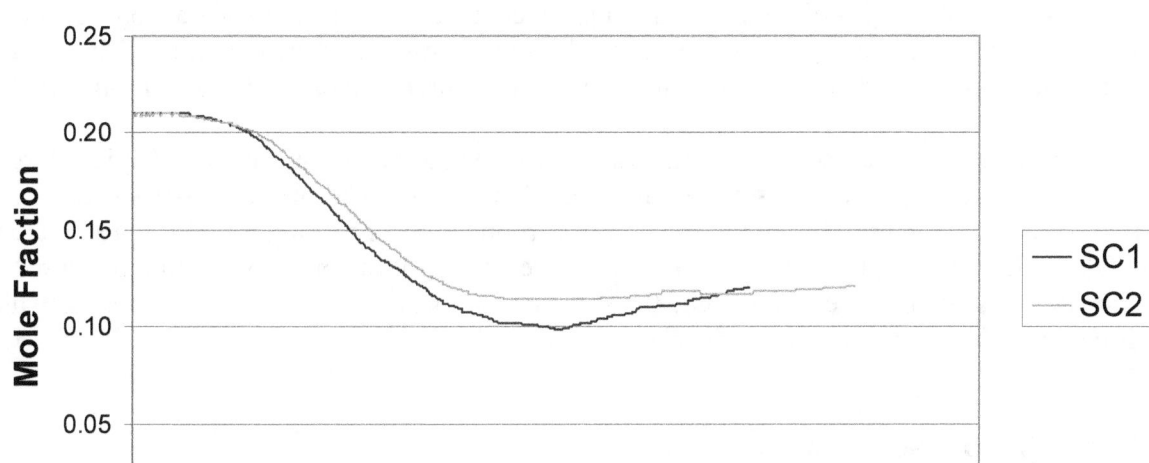

E. Species Losses During Transport

Loss of a combustion product during travel down the corridor is quantified by the ratio of the concentration at location 2 to location 4 relative to the same ratio for CO_2. Analysis of the data in Tables 13 and 14 produces the results in Table 24:

Table 24. Ratios of Downstream (Location 4) to Upstream (Location 2) Concentrations (Post-flashover Data)

	Analyzer	Sofa	Bookcase	Cable	Bookcase/PVC
CO_2	NDIR	0.65 ± 0.05	0.48 ± 0.05	0.58 ± 0.01	0.60 ± 0.04
CO	NDIR	0.54 ± 0.21	0.41 ± 0.09	0.57 ± 0.02	0.75 ± 0.26
CO_2	FTIR	0.41 ± 0.14	0.73	0.35 ± 0.02	0.25 ± 0.24
CO	FTIR	0.07 ± 0.04	0.57	0.12 ± 0.01	0.10 ± 0.02
HCl	FTIR	0.08 ± 0.05	0.53 ± 0.50	0.21 ± 0.02	0.11 ± 0.09
HCN	FTIR	0.17 ± 0.09	0.43	0.45 ± 0.02	0.39 ± 0.18
Smoke	Filter	0.47 ± 0.10	0.45 ± 0.22	0.18 ± 0.04	0.39 ± 0.18

F. Estimates of Toxic Gas Yields with Uncertainties

Table 25 contains the yields of the combustion products calculated using the data from location 2. The estimated uncertainties reflect the repeatability of similar tests, discounting of disparate individual test results, and degree of proximity of the measured values to the background levels.

It was noticed (and is discussed in the following Section) that the post-flashover CO yields are lower than expected. For that reason we have compiled in Table 26, for the post-flashover portion of each test, an estimate of the fraction of carbon atoms appearing in CO (approximated as $[CO]/\{[CO] + [CO_2]\}$) at both location 1 (near the ceiling of the burn room) and location 2 (outside the burn room). We also include the values from location 3 to assess the extent to which the fire plume chemistry has been quenched.

G. Archiving of Test Data

All of the raw data (*ca.* 130 instruments, thousands of readings per instrument) from the tests reported here are to be available in a companion report.[33] This will be in the form of spreadsheets and graphs.

Table 25. Yields of Combustion Products Calculated from Location 2 Data

Gas	Fire Stage	Sofa	Bookcase	PVC Sheet	Cable
CO_2	Pre-flashover	$1.59 \pm 25\%$	$0.50 \pm 50\%$	--	$0.120 \pm 45\%$
	Post-flashover	$1.13 \pm 25\%$	$1.89 \pm 75\%$	--	$1.38 \pm 15\%$
CO	Pre-flashover	$1.44 \times 10^{-2} \pm 35\%$	$2.4 \times 10^{-2} \pm 55\%$	--	$5.5 \times 10^{-3} \pm 50\%$
	Post-flashover	$5.1 \times 10^{-2} \pm 25\%$	$4.6 \times 10^{-2} \pm 30\%$	--	$1.48 \times 10^{-1} \pm 15\%$
HCN	Pre-flashover	$3.5 \times 10^{-3} \pm 50\%$	$4.6 \times 10^{-4} \pm 10\%$	--	$6.3 \times 10^{-4} \pm 50\%$
	Post-flashover	$1.5 \times 10^{-2} \pm 25\%$	$2.5 \times 10^{-3} \pm 45\%$	--	$4.0 \times 10^{-3} \pm 30\%$
HCl	Pre-flashover	$1.8 \times 10^{-2} \pm 30\%$	$2.2 \times 10^{-3} \pm 75\%$	--	$6.6 \times 10^{-3} \pm 35\%$
	Post-flashover	$6.0 \times 10^{-3} \pm 35\%$	$2.2 \times 10^{-3} \pm 65\%$	$2.3 \pm 85\%$	$2.1 \times 10^{-1} \pm 15\%$
NO_2	Pre-flashover	$<7 \times 10^{-2}$	$<2 \times 10^{-2}$	--	$<4 \times 10^{-3}$
	Post-flashover	$<1 \times 10^{-3}$	$<1 \times 10^{-3}$	--	$<1 \times 10^{-3}$
Acrolein	Pre-flashover	$<8 \times 10^{-3}$	$<2 \times 10^{-3}$	--	$<4 \times 10^{-4}$
	Post-flashover	$<1 \times 10^{-4}$	$<1 \times 10^{-4}$	--	$<1 \times 10^{-4}$
Formaldehyde	Pre-flashover	$<2 \times 10^{-2}$	$<2 \times 10^{-3}$	--	$<8 \times 10^{-4}$
	Post-flashover	$<8 \times 10^{-4}$	$<4 \times 10^{-4}$	--	$<7 \times 10^{-4}$

Table 26. Fraction of Combustible Carbon Appearing in Carbon Monoxide (NDIR, Post-flashover Data)

Test	Location 1	Location 2	Location 3
SW10	0.040	0.077	0.076
SW11	0.034	0.069	
SW12	0.037	0.080	0.071
SW13	0.016	0.064	0.056
SW14	0.029	0.072	0.070
BW3	0.061	0.091	0.100
BW4	0.075	0.085	0.069
BW6	0.077	0.093	0.037
BP1	0.080	0.100	0.102
BP2	0.125	0.068	0.060
PQ1	0.084	0.152	
PQ2	0.073	0.143	
PW1	0.101	0.151	0.147
PW2		0.151	0.145

V. DISCUSSION

A. Overall Test Quality

The most important outcome of this series of tests is a reliable, well-documented set of combustion product yields. This includes the numerical values themselves, the specific combustion conditions under which they were obtained, the uncertainty in their calculated values, and the repeatability of the tests.

Next, most important is the initiation of the development of a standard protocol for obtaining yield values from a wider variety of test specimens. This includes test conduct procedures, experimental design, instrumentation, species sampling, and data reduction.

Third, it is important to evaluate the quality of the derived knowledge in the context of its intended use. The yield information would be used with a fire model (zone or CFD) to generate the time-dependent environment generated by a fire. Equations such as those in ISO/TS 13571[6] would then be used to assess whether the combination of occupancy design, contained combustibles, and occupant/responder characteristics lead to the desired level of life safety.

The documentation has been provided in the earlier sections. The following examines the context and quality of the results.

B. Test Repeatability

Even under controlled laboratory conditions, attaining a reasonable degree of consistency in replicate fire tests requires both conceptual understanding of the phenomena and attention to detail. Even then, much of the success is attributed to art as well as science and engineering. In the current series, two to five tests of each of the complex fuels led to an appraisal of how likely the results from a single test might be representative. The results were within the range needed for the intended application.

By all measures, the repeatability of the sofa tests SW10 through SW14 is excellent. From Figure 15, there is qualitative agreement of the shapes of the mass burning rate curves. Table 17 shows that the global equivalence ratios for the tests are also similar. Table 22a shows that the variability in the post-flashover yields of CO_2, CO, and HCN are within ± 25 %, while the variability in the yield of the more difficult-to-sample HCl is ± 35 %. The pre-flashover yield values, for reasons discussed below, are repeatable to within a factor of two.

Table 22b shows that the pre-flashover yield values are of comparable repeatability, although the uncertainty in the HCN result is somewhat higher.

For sofa tests SW1 through SW3, which did not reach flashover, the mass burning rate curves (Figure 16) are similar and the later pre-flashover burning data (Table 22b) show the CO_2, CO and HCl yields all repeatable to within ± 36 % and the HCN yield repeatable to within ± 45 %.

The final yields (Table 16) from the two closed room sofa tests (SC1 and SC2) are repeatable to within ± 20 %.

The results of the four cable tests (PQ1, PQ2, PW1, PW2) indicate qualitatively similar results (Table 25). Post-flashover yield repeatability is typically ± 15 % to 30 %, with the pre-flashover repeatability somewhat higher, but within a factor of two.

For the four bookcase tests (BW3, BW4, BW6, BW7) in which NDIR data were obtained, the repeatability was not as good as for the other combustibles. The post-flashover and pre-flashover yield repeatability values for CO_2 are *ca.* ± 75 % and ± 30 %, respectively; the CO values are *ca.* ± 30 % and ± 55 %. There were only two bookcase tests for which we obtained FTIR data. Fortunately, the yield agreement was good. For HCN, the post-flashover and pre-flashover yield repeatability values are *ca.* ± 45 % and 10 %, respectively. For HCl, they are 65 % and 75 %.

The post-flashover HCl yields from the three PVC sheet tests spanned over an order of magnitude. See the discussion below.

In summary, the repeatability of the yields values obtained here for three of the combustibles is sufficient for determination of whether a bench-scale apparatus is producing results that are similar to or different from the real-scale results here. The PVC sheet, from which only HCl yield data could be obtained, can only provide an indicator of appropriateness and then only for post-flashover simulation.

The repeatability results indicate an uncertainty in the fractional effective dose (FED) calculations in ISO/TS 13571 that is comparable to the uncertainty in the equations themselves. This is especially so since a large fraction of fire deaths result from post-flashover fires (reducing the importance of the larger variance in the pre-flashover gas yields) and since CO is always a major (if not the dominant) incapacitating toxicant (reducing the importance of the variance in the other toxicants).

C. Species Sampling and Measurement

1. CO$_2$ and CO

The CO_2 and CO data have low variability and good consistency between analyzers at location 2 during the post-flashover period. The CO_2 concentrations are high (*ca.* 5 % to 10 % by volume) and well above the sensitivity limit of the analyzers, and the sampling is reliably from the well established upper layer. Appropriately, the scatter among replicate tests is about 20 %, the lowest for any set of concentration measurements and not far from the estimated repeatability of the mass burning rate. The CO concentrations are also high (0.5 % to 1 % by volume) and show repeatability similar to the CO_2 results. As indicated in Section IV.C.1, the FTIR and NDIR instruments show good agreement for both gases.

The same was true at location 2 during the late pre-flashover period in tests SW1 through SW3, despite some of the concentrations being *ca.* an order of magnitude lower than after flashover. However, during the *general* pre-flashover burning periods for all tests, including the earlier pre-flashover periods in tests SW1 through SW3, distinctly higher variability was observed. Concentrations were yet another order of magnitude lower. Care was needed to adjust the sampling time period such that the thermocline showed the probe tip was sampling from the upper layer. Turbulence in the effluent stream was likely to result in non-uniform mixing with lower layer air. The burning rates of replicate tests varied more during the early growth period than later on.

An additional observation is that the location 2 pre-flashover measurements using the FTIR analyzer were consistently smaller than those using NDIR for both gases (Table 19) and that the ratio varies widely from test to test. The exceptions are the data from the late (non-flashover) portions of tests SW1 through SW3. This suggests that it took more time to establish the upper layer than was typically available before the transition to flashover occurred. However, there is no hard evidence to verify this hypothesis.

At location 4 during the post-flashover period, the upper layer visually encompassed the probe tips, and the variation in CO_2 and CO concentrations was similar to that at location 2 for the NDIR measurements. The FTIR concentration values for CO_2 and CO were again lower than the NDIR values.

The concentrations of species at location 4 during the pre-flashover periods were very low, with all but the CO_2 values being near the detection limits of the analyzers. The replicate tests SW10 through SW14 show some small degree of consistency.

Each cell in Table 16 represents the volume fraction or yield integrated from the beginning of the test to that point in time. The values are probably somewhat high, since the sampling was performed at only one point near the top of the upper layer, and it was likely that there was a decreasing concentration gradient from the ceiling downward. The early NDIR yields for CO_2 and CO are remarkably close to those for the open door sofa tests (Tables 15 and 25). As the fire progressed, the CO_2 yield decreases and the CO yield increases, as expected from burning in an increasingly vitiated atmosphere.

2. HCl and HCN

The concentrations of these gases were only measured using FTIR. Thus, the same sampling considerations that were discussed above apply here. To mitigate these effects, we use the ratios of the concentrations and yields of HCl and HCN to the corresponding values for CO_2.

From comparison of the sensitivity limits in Table 9 and the measured concentrations in Tables 13 and 14, one can see that there are some combinations of location, fire phase, and combustible for which the measurements are very close to the background. The location 2 data are sufficient to obtain reasonable post-flashover yield values and pre-flashover yield estimates for all three principal combustibles.

The HCl concentration data from the PVC sheet tests are high enough to obtain post-flashover HCl yields. Table 21 shows that the yield values from the multiple tests have a high degree of scatter and that the yields are at least as high as the notional yield. This most likely indicates that the HCl is being pyrolyzed from the test specimen faster than the carbon-containing species. It is not likely that this is due to an artifact of the HCl measurement for two reasons. First, the FTIR instruments were carefully calibrated. Second, if the FTIR spectra were indicating high yield values, the CO and CO_2 results should also be high. Table 19 shows that for the bookcase/PVC sheet tests, very high HCl yield values would be inconsistent with the more modest deviations from unity in the yields of these gases between the two types of instruments.

The post-flashover location 4 measurements for HCl and HCN are high enough to obtain estimates of the degree of loss of the compounds down the length of the corridor. By contrast, nearly all the pre-flashover values can be expected to have a high degree of uncertainty and are not useful for even a rough estimate of losses down the corridor.

The two closed-room sofa tests, SC1 and SC2, produced very repeatable yield results for these two gases, the exception being a burst of HCl early in SC2.

3. Other Gases

The equations in ISO/TS 13571 include additional gases to be included in estimating the time available for escape or refuge from a fire: HBr, HF, SO_2, NO_2, acrolein (C_3H_4O) and formaldehyde (H_2CO). There was no Br, F, or S in any of the products examined in this project, so the first three of these gases were not expected. The presence of the latter three was not detected, thus establishing the upper limits of their presence at 100, 10, and 50×10^{-6} volume fraction, respectively.

All three of these gases are sensory irritants. Their incapacitation concentrations from ISO/TS 13571, their ratios normalized to HCl, and the concentration (location 2, post-flashover) ratios of the gases in this study are shown in Table 27. The measured pre-flashover concentrations were too low to obtain usable comparison.

Table 27. Limits of Importance of Undetected Toxicants

	Volume fraction x 10^6				Ratio to [HCl]			
	HCl	NO_2	C_3H_4O	H_2CO	HCl	NO_2	C_3H_4O	H_2CO
Incapacitating level	1000	250	30	250	1.00	0.25	0.030	0.25
Sofa	800	<100	<10	< 50	1.00	< 0.12	< 0.012	< 0.06
Bookcase	20-200	<100	<10	< 50	1.00	< 5 − 0.5	< 0.5 − 0.05	< 2.5 − 0.25
Cable	1400	<100	<10	< 50	1.00	< 0.007	< 0.0007	< 0.04
PVC sheet	8000	<100	<10	< 50	1.00	< 0.012	< 0.0012	< 0.006

From this analysis, the maximum concentrations of NO_2, formaldehyde and acrolein that could have been present would have had secondary contributions to incapacitation relative to the concentration of HCl in the sofa, cable and PVC sheet tests. In the bookcase tests, where the HCl levels are low, the other irritants could be important. However, the high levels of CO in those tests suggest a secondary role for the irritant gases in causing incapacitation.

Levin *et al.* have developed extensive information on the effects of gas mixtures on rat lethality and incapacitation[34]. They used those data to test whether the toxic potency of a small number of gases could account for the lethality of the effluent from a variety of materials. The apparatus conditions were typical of pre-flashover combustion. Within the uncertainty in the results, $\pm$ 30 %, there as no need to invoke additional toxicants. Combined with the results obtained here for post-flashover conditions, it suggests that a set of upper limit criteria for these gases would be a reasonable criterion for the accuracy of a bench-scale apparatus.

4. Species Measurement Using FTIR Spectroscopy

There is considerable interest in adding FTIR spectroscopic analysis to fire test apparatus. A major European program[35] developed extensive information on the technique, and there are documents under development in ISO TC92 SC1 and SC3 to standardize the implementation.

We were able to obtain usable information using this technique. There are a number of lessons emerging from this test series that can provide useful input to these efforts, such as the following:

- The application of FTIR spectroscopy to fire testing requires the constant attention of an experienced professional at a level well beyond the demands of the more traditional fire test instrumentation.

- To maximize the opportunity for obtaining time dependent concentration data, we selected a small volume cell of short optical path length and operated without a soot filter. While some cleaning was necessary, it was not a major impediment. The short path length did limit the sensitivity, but did not seriously compromise our ability to determine toxicologically important levels of the major gases, as noted above.

- The long, heated lines used here (and recommended in the SAFIR report[35]) enabled quantitative collection of HCl, a compound that is generally regarded as difficult to determine.

D. Combustion Conditions

The contents of Table 18 show that the combustion efficiency declines slightly as the fire proceed through flashover. As is to be expected, the $[CO]/[CO_2]$ ratio increases.

The global equivalence ratios are surprising. The local equivalence ratios during ventilation limited (post-flashover) combustion should be above unity. However, other research is now indicating that two thirds or more of the air entering a flashed over compartment is forced out of the compartment before approaching the fire zone.[36]

E. Loss of Acid Gases During Transport

In light of the above discussion, we used only post-flashover data in the estimation of the degree to which HCl and HCN were lost to walls or deposited on smoke aerosol. In interpreting the data, one must recognize that the cells in Table 24 often reflect a small number of tests.

The first observation is that the NDIR data indicate about a factor of two dilution of these two fixed gases (CO and CO_2) with air from the lower layer for all combustibles. This is a reasonable finding since the dilution both of these gases should be affected only by the traverse time down the corridor, and that should be uniform since the doorway flows are uniform (Table 14A).

From there, the picture becomes more complex. The CO_2 data from the FTIR analyzers give results that are similar to the NDIR data. However, the data for the other effluent components show distinct dependence on the combustible.

- The FTIR CO concentration changes differ significantly from the NDIR changes for the sofas and cable.

- The CO, HCl and HCN concentrations generally decrease by similar factors for a given combustible, even though the factors are combustible-dependent. [The exception to this is HCN from the cable and bookcase/PVC fires, the reason for which is also unknown.] The bookcase results are similar to the dilution of CO_2. The sofa, bookcase/PVC and cable factors are more severe.

The cause of this is not understood. However, soot particles and aqueous aerosols are characterized by their number density, surface area, and hydrophilia. In these tests, only the soot *mass* was measured. It may well be that the smoke from the sofa, PVC and cable materials have a greater affinity for acid gases and CO than does the smoke from the bookcases.

The inference we draw from these results is that for large fires of some combustibles, there can be little loss of reactive gases. This is consistent with a previously reported analysis of other experimental data.[7] However, for some other combustibles, loss factors of two to five beyond dilution are possible. Care should be taken not to extend these limited findings to other commercial products. In the absence of a comprehensive study of the relationship between smoke character and gas absorption, safety engineers are most likely to continue to assume there is no loss of toxicants, the more conservative approach.

F. Yield Values

During vigorous combustion, the yields of CO_2 and HCl should approach their notional values. As can be seen from Table 21, the post-flashover values of CO_2 from all three combustibles do just that, given the conversion of up to *ca.* 10 % of the carbon to carbonaceous smoke and CO and some formation of carbonaceous char residue. Under pre-flashover conditions, the CO_2 fraction from the sofas was as large as expected. However, for unknown reasons the bookcase and cable fractions were far lower. In the closed room sofa tests, the yield begins at about the notional level, then declines to about half that as room vitiation affects the completeness of combustion.

The HCl yields are close to notional under post-flashover conditions for the sofas, bookcases, and cable arrays. By contrast, in prior work[37], *ca.* 40 % or less of the HCl from room-scale tests reached the analyzers. We attribute the present improvement to the use of calcium silicate walls (rather than drywall), the use of heated transfer lines to the FTIR spectrometers, and the absence of soot filters in those transfer lines. The somewhat lower value for the power cable may reflect the known HCl reaction with the calcium carbonate filler in the cable jacket.

The pre-flashover stages of the sofa tests and the post-flashover stages of the PVC/bookcase tests produced HCl yields well above the notional values. The cause of these is unexplained. However, in both cases, relatively little of the specimen mass was volatilized, and it is possible that a disproportionate fraction was fire retardant (sofa tests) or HCl (PVC tests).

Generally little of the nitrogen in the combustibles ends up in HCN. The observed exceptions are with the urea formaldehyde resin in the bookcase particle board and the nylon in the wire insulation, where over 10 % of the N atoms appear in HCN. This is consistent with the results of prior room-scale tests with a different urethane foam[37] in which 5 % to 10 % of the fuel nitrogen appeared as HCN and where (as noted above) the sampling of reactive gases was less efficient.

It is interesting to note that the addition of the PVC sheet to the bookcase fires leads to an order of magnitude increase in the yield of HCN (Table 15). Perhaps the flame inhibition by the chlorine atoms is reducing the ability of the flame radicals to oxidize the HCN to one of the nitrogen oxides. Otherwise, Table 21 shows only modest differences in the conversion of fuel nitrogen to HCN despite large differences in the chlorine content of the fuel.

The early NDIR yields in the closed-room tests (Table 16) are consistent with the pre-flashover yields from the equivalent (later pre-flashover) phase of open room tests SW1 through SW3 (Table 15). [Note that the first entries in Table 16 occur at 150 s; prior to that, the mass loss and concentration values were too small and noisy to obtain reliable ratios.] As noted above, the FTIR-derived CO and CO_2 yields are consistently somewhat lower than their NDIR-derived counterparts. However, the ratios of the yields of HCN and HCl to the FTIR-derived CO_2 yield are a fair basis for comparison and are similar to the later pre-flashover phase of tests SW1 through SW3. Because the mass burning rate falls sharply as the oxygen concentration drops, the yields from SW1 through SW3 can be used to approximate the overall yields from the closed room tests. The volume fraction section of Table 16 does not show evidence of large increases in the rates of generation of HCN or CO as the fires approach extinction.

Of most interest are the post-flashover yields of CO. A number of room-scale fire studies have indicated that the yield of CO is approximately 0.2 (g CO/g fuel consumed) and that this value is not very dependent on the combustible.[9] In this study, the post-flashover CO yields from the cable fires approach this, with a mean of *ca.* 0.15 g/g. The sofas and bookcases appear to generate about one quarter of the expected value.

A first consideration is whether the tests truly reached flashover. Examination of the test videos and data logs indicate that each "declaration of flashover" occurred when key characteristics were observed: significant oxygen depletion within the fire room, high temperature in the upper layer of the burn room and in the upper portion of the doorway, flames out the doorway.

Experimental errors of a sufficient magnitude are highly unlikely. Since two different types of analyzers with independent sampling lines produced comparable CO yields, the difference from the expected value cannot be attributed to an instrumental or sampling error. Since the same calculations produced CO_2 yields near the notional limits, there cannot be a missing factor in the data reduction.

It is possible that a large amount of CO is formed in the room, but is oxidized in the secondary burning at the doorway. Computer simulations of room fires using FDS indicate that the environment at Location 1 should be highly vitiated and that the CO should be at its peak there. The test records indicate low oxygen levels. However, Table 26 shows that the fraction of fuel carbon appearing in CO is actually lower at Location 1 than at Location 2. It is possible that at Location 1, a sizable fraction of the carbon exists as uncombusted pyrolyzate, some of which is partially oxidized to CO in the doorway. This would account for the observations of relative CO concentration in the two locations. However, this is speculative and at present, there is no firm explanation for this behavior.

A more likely hypothesis is as follows. Large quantities of pyrolyzate are generated during flashover. Much of these consume the limited available oxygen, forming CO, but leaving much of the organic matter unoxidized. As these gases reach the doorway and begin to entrain fresh air, more of the organic matter is oxidized to CO. Some of the CO is also oxidized to CO_2. Combined, these processes set up a dynamic situation where the observed [CO]/[CO$_2$] ratio and the yield of CO depend on the degree of air-effluent mixing and the rate of cooling of the total flow.

Different fires and different stages of those fires are likely to be accompanied by differing degrees of CO formation and burnout. Thus, we suggest that for fire hazard and risk assessments, one should use the CO yield value of 0.2 g CO per g fuel consumed. Bench-scale combustors typically used for generating toxic potency data generally do not have the potential for the secondary combustion processes described above. Thus, for assessing the accuracy of the data from such apparatus, it is also appropriate to use the CO yield value of 0.2 g CO per g fuel consumed.

G. Use of the Results

The yield data developed here are ready for use in determining whether and how to use a bench-scale apparatus for generating toxic potency data of known accuracy. Generically, the following steps are suggested:

- Combust samples of these specimens in the bench-scale device under a range of combustion conditions appropriate for well-ventilated and underventilated fires.

- Determine the degree to which agreement is reached with the yields measured here.

- For the gases whose yields here were below the detection limits, determine whether the bench-scale results are consistent with these detection limits.

- For CO, keep in mind that other studies have measured yields significantly larger than the values determined here and use the CO yield value of 0.2 g CO per g fuel consumed.

- Appropriate weighting of the comparisons for individual gases can be derived using the equations in ISO/TS 13571.

VI. CONCLUSION

It is important to be able to demonstrate how well a bench-scale toxic potency measurement apparatus reflects the effluent produced in real fires of the same combustible. This report documents the measurement of the yields of the prime toxicants (CO_2, CO, HCl, HCN) and smoke from the combustion of three complex products (sofa cushions, bookcases, and power cable) in a room connected to a long corridor. There are results for both the pre-flashover and post-flashover stages of the fires, with additional post-flashover yield data on a PVC material.

The repeatability of the yields values obtained here is sufficient for determination of whether a bench-scale apparatus is producing results that are similar to or different from the real-scale results here.

The uncertainty in the post-flashover data is smaller due to the larger species concentrations and the more fully established upper layer from which the fire effluent was sampled. The toxicant yields from sofa cushion fires in a closed room were similar to those from pre-flashover fires of the same cushions in a room with the door open.

Since a large fraction of fire deaths result from post-flashover fires and since CO is always a major (if not the dominant) incapacitating toxicant, the repeatability results indicate an uncertainty in the fractional effective dose (FED) calculations that is comparable to the uncertainty in the equations themselves. The repeatability values should also be sufficient to determine whether a bench-scale apparatus is producing results that are similar to or different from the real-scale results here.

Other toxicants (NO_2, formaldehyde and acrolein) were not found. Concentrations below the detection limits were shown to be of limited toxicological importance relative to the detected toxicants.

The use of Fourier transform infrared (FTIR) spectroscopy was shown to be a useful tool for obtaining toxicant concentration data. However, its operation and interpretation are far from routine. The agreement between the FTIR instruments and conventional non-dispersive infrared analyzers was not perfect but was reasonable enough to identify situations where the effluent sampling may have been compromised and where signals were approaching the background limits.

Measurements at both ends of the corridor provided an indication of the degree to which the combustion product concentrations decreased relative to simple dilution. The losses of CO, HCN, and HCl were found to be dependent on the combustible. The downstream to upstream concentration ratios varied from unity for some fuels to a factor of five smaller for others.

The yield of CO for the sofa and bookcase tests was significantly lower than the expected value of 0.2, while the CO yield for the cable tests was close. The determinations were shown to be accurate. It is suggested that one should use the CO yield value of 0.2 g CO per g fuel consumed for both fire safety analyses and for assessing the accuracy of bench-scale combustors for generating toxic potency data

VII. ACKNOWLEDGMENTS

The authors express their appreciation for the contributions of many people in bringing success to such an expansive project. George Mulholland designed the smoke sampling apparatus, which was built and operated by Michael Selepak; Dr. Mulholland and Jenny Oran analyzed the data. David Stroup operated the large-scale test facility and with Lauren DeLauter and Gary Roadarmel built the room/corridor assembly and assisted in the tests. Richard Harris designed the wet chemistry apparatus and applied it to many of the tests, assisted by Walid Awad. Jack Lee assisted in numerous ways during the room construction and test conduct. Jonathan Demarest also assisted with the tests and in reducing the data. Dr. Kurt Reimann of BASF provided information helpful in the design of the sofas; Douglas Wetzig of PolyOne supplied the PVC sheet; David Mercier of Southwire, Inc. provided useful information on the formulation of the cable. Schwartzkopf Laboratories performed the elemental and thermal analyses of the combustibles.

The research was co-sponsored by the Alliance for the Polyurethane Industry, the American Plastics Council, DuPont, Lamson & Sessions, Underwriters Laboratories, and the Vinyl Institute under the aegis of the Fire Protection Research Foundation and the supervision of Frederick Mulhaupt and Steven Hanly.

APPENDIX A. GRAPHS OF TEST DATA

Figures A1a, A1b. Data from Test SW1, Location 2

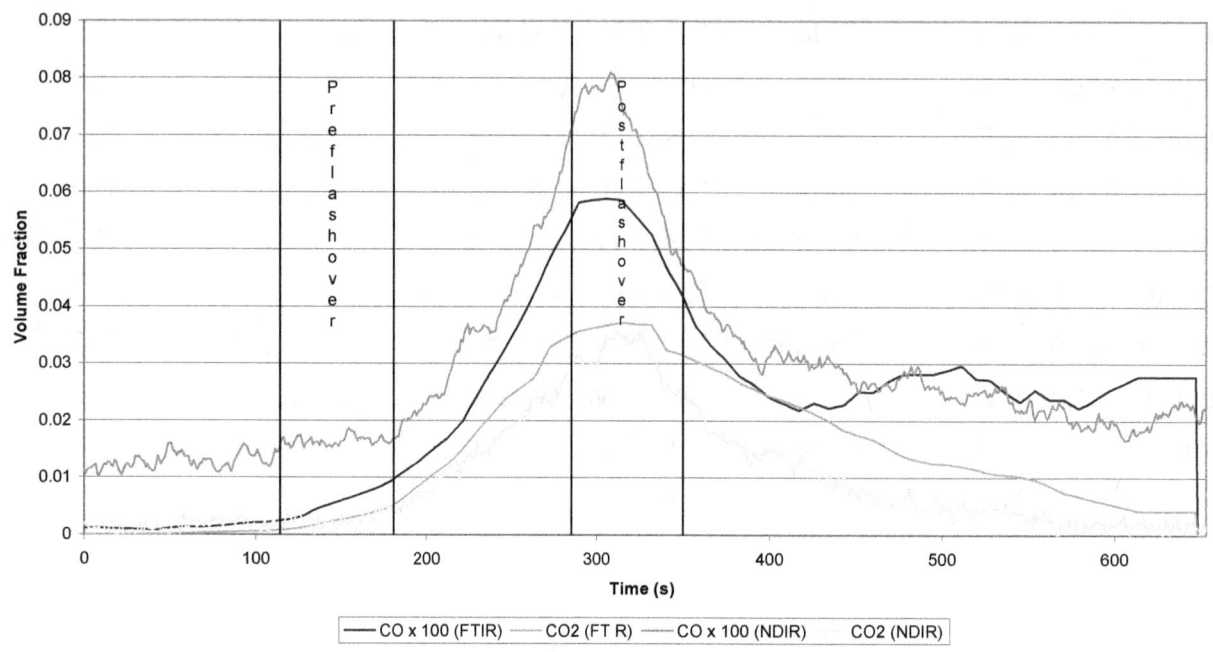

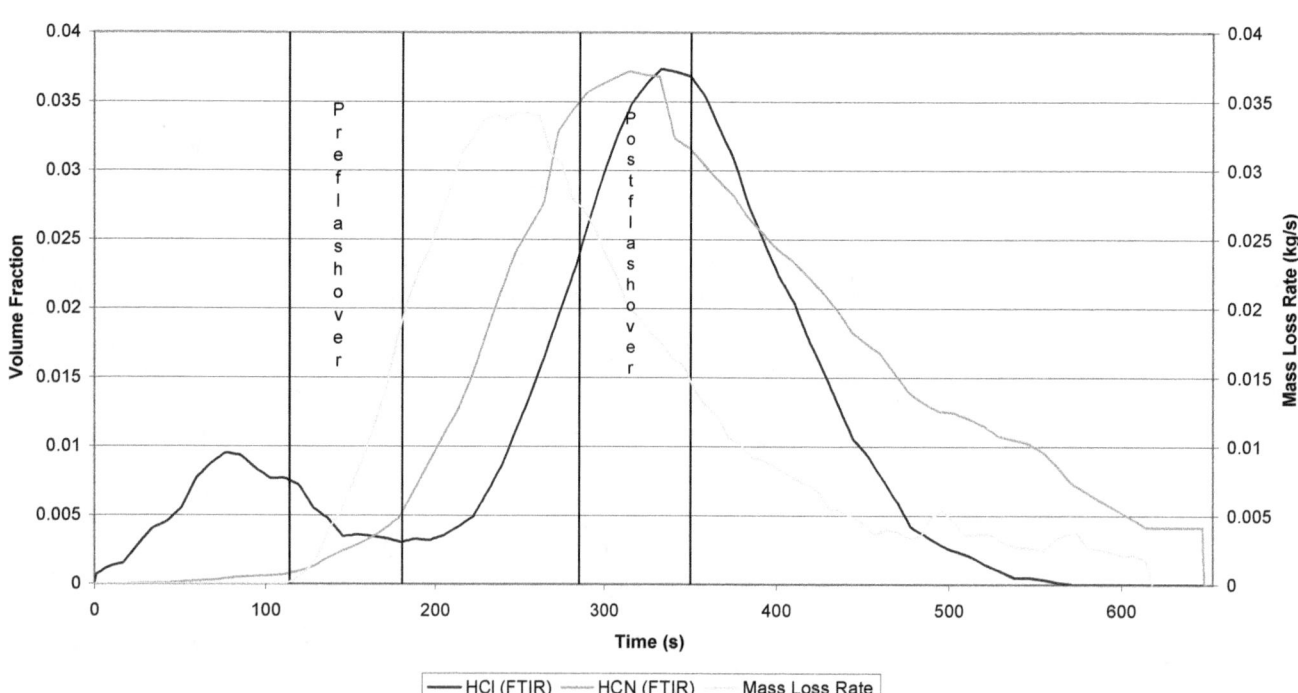

Figures A1c, A1d. Data from Test SW1, Location 4

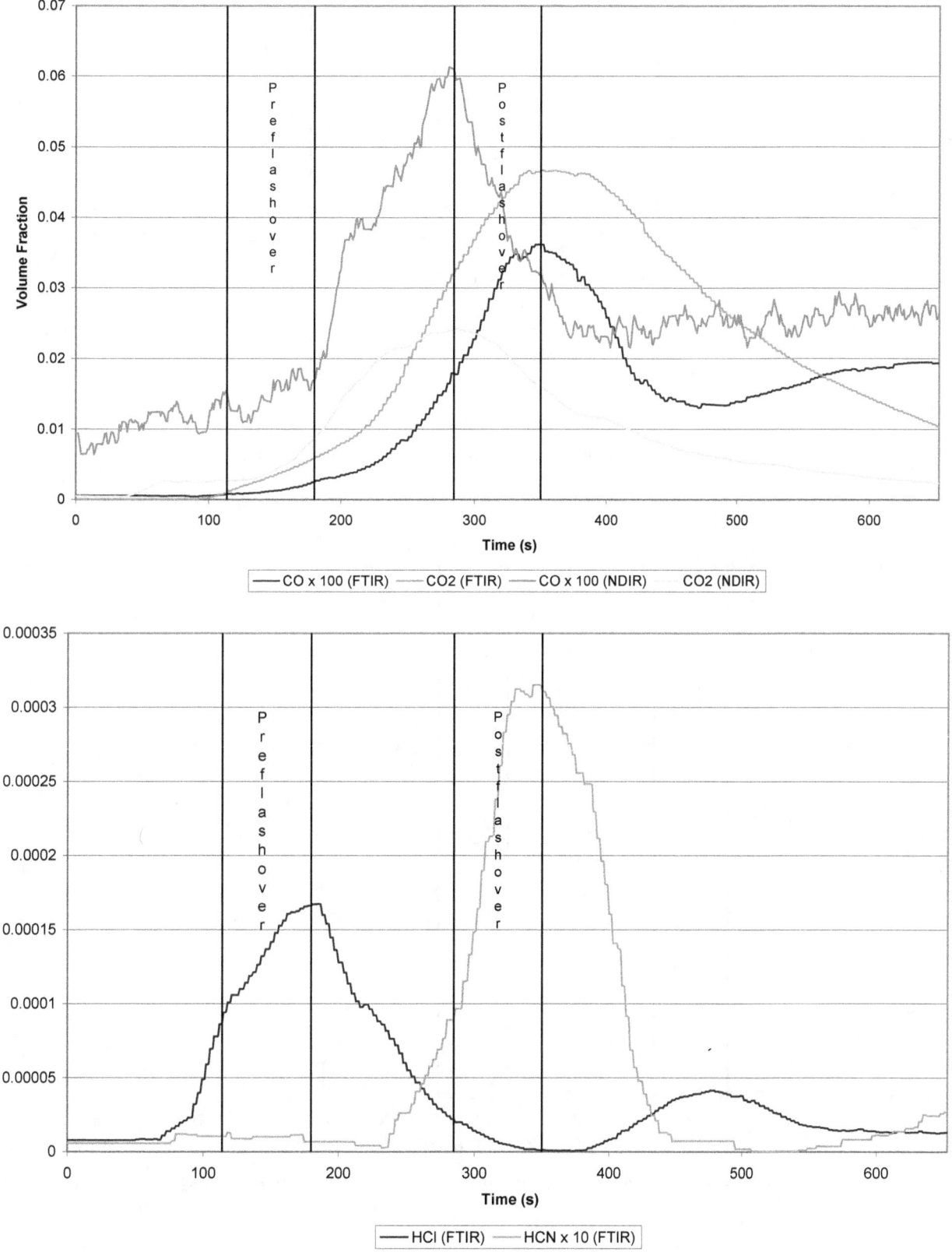

95

Figures A2a, A2b. Data from Test SW2, Location 2

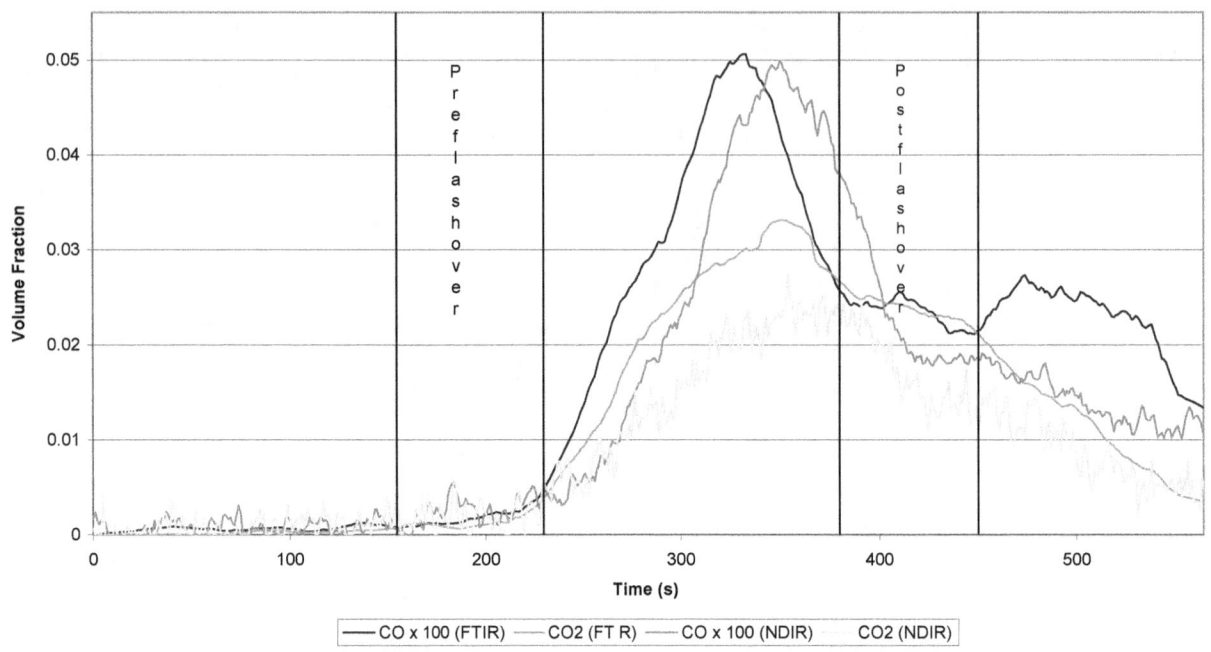

CO x 100 (FTIR) — CO2 (FT R) — CO x 100 (NDIR)   CO2 (NDIR)

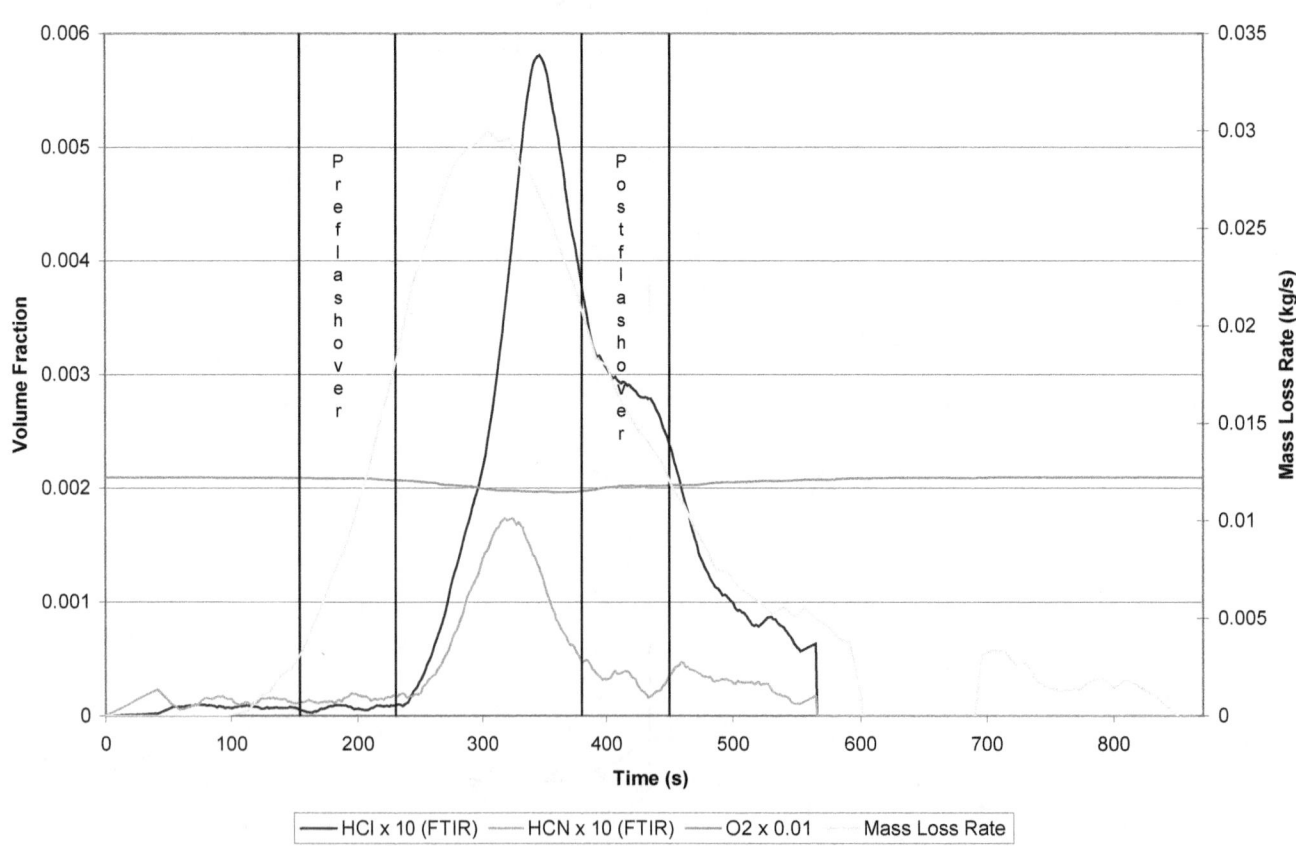

HCl x 10 (FTIR) — HCN x 10 (FTIR) — O2 x 0.01   Mass Loss Rate

Figures A2c, A2d. Data from Test SW2, Location 4

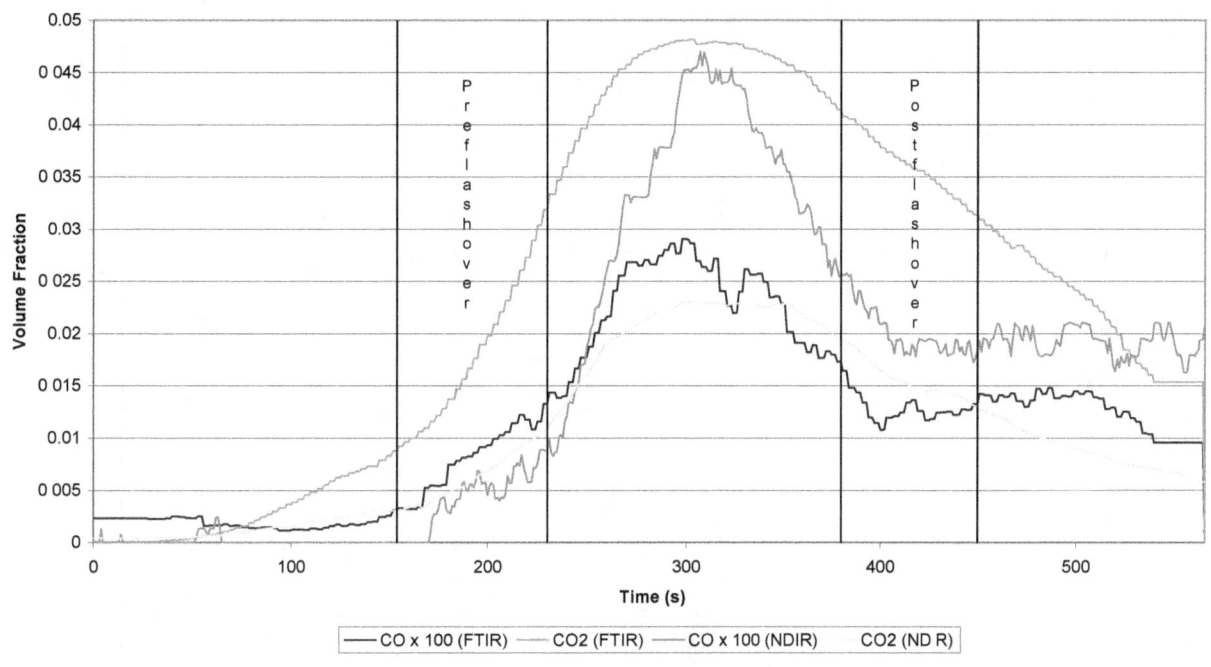

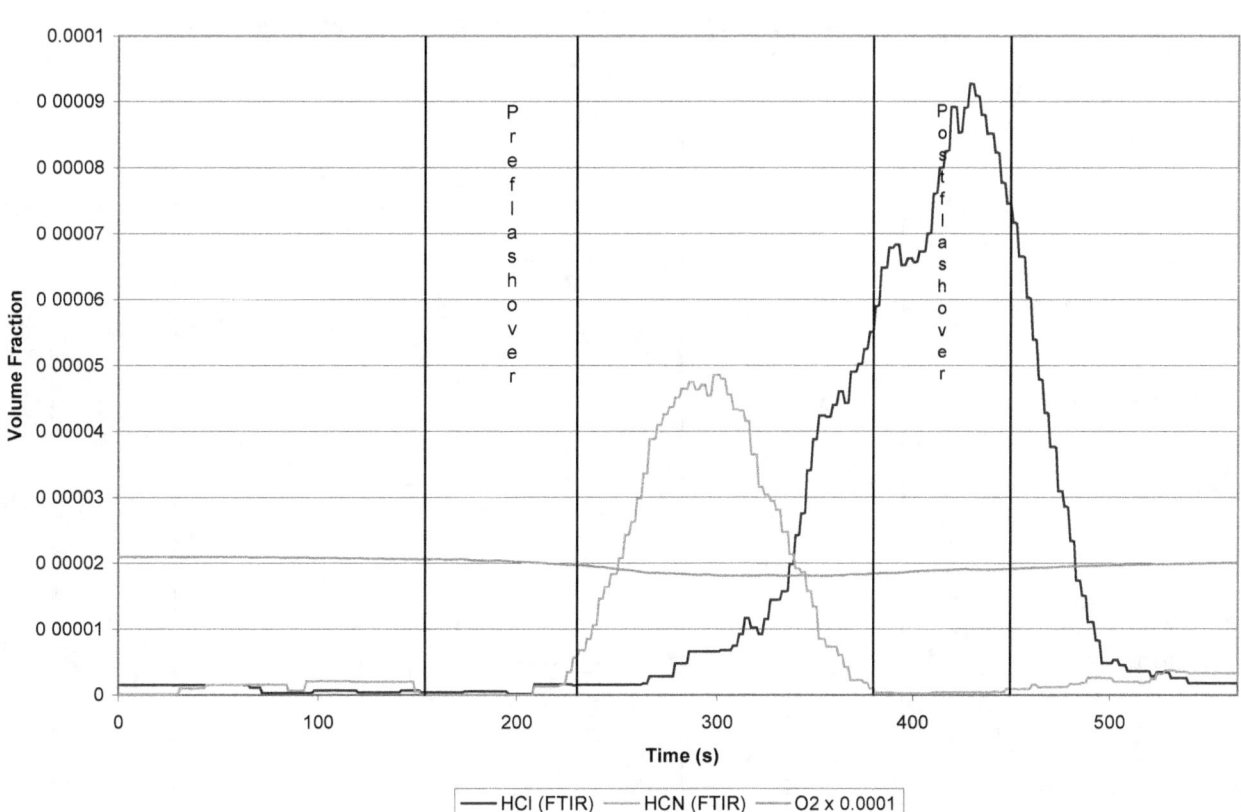

Figures A3a, A3b. Data from Test SW3, Location 2

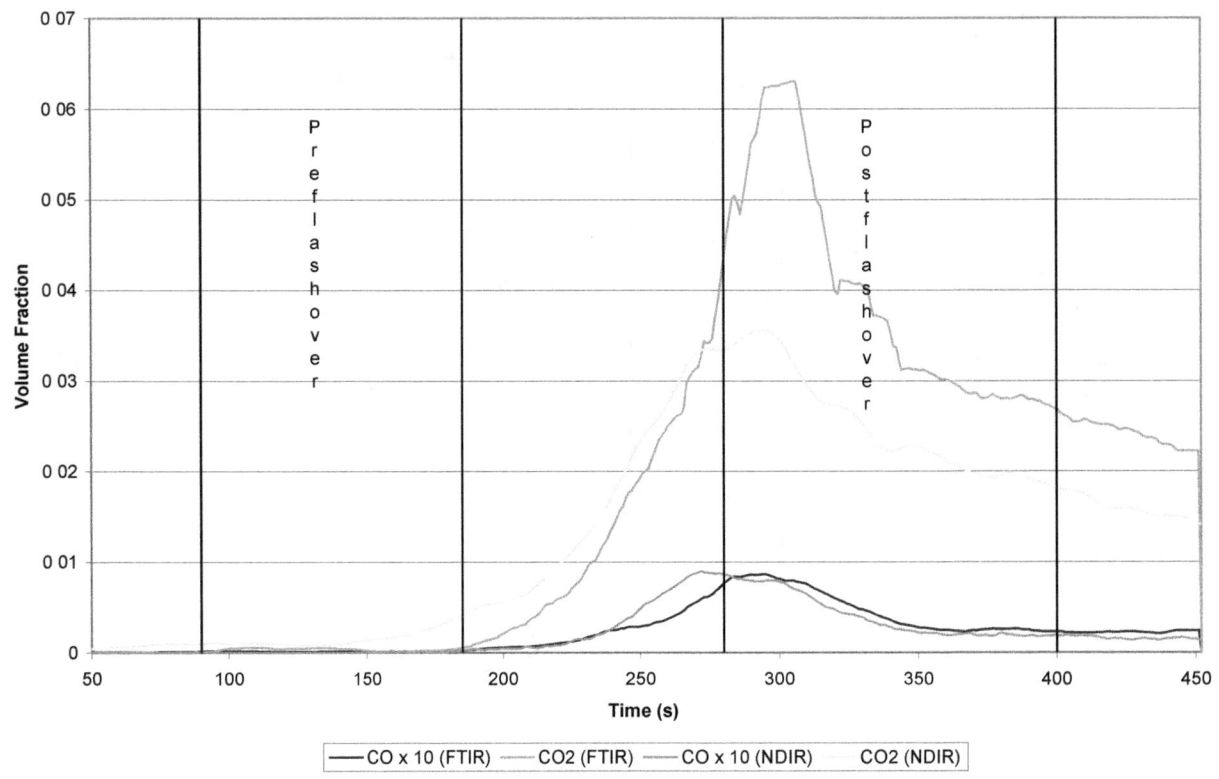

CO x 10 (FTIR) —— CO2 (FTIR) —— CO x 10 (NDIR) —— CO2 (NDIR)

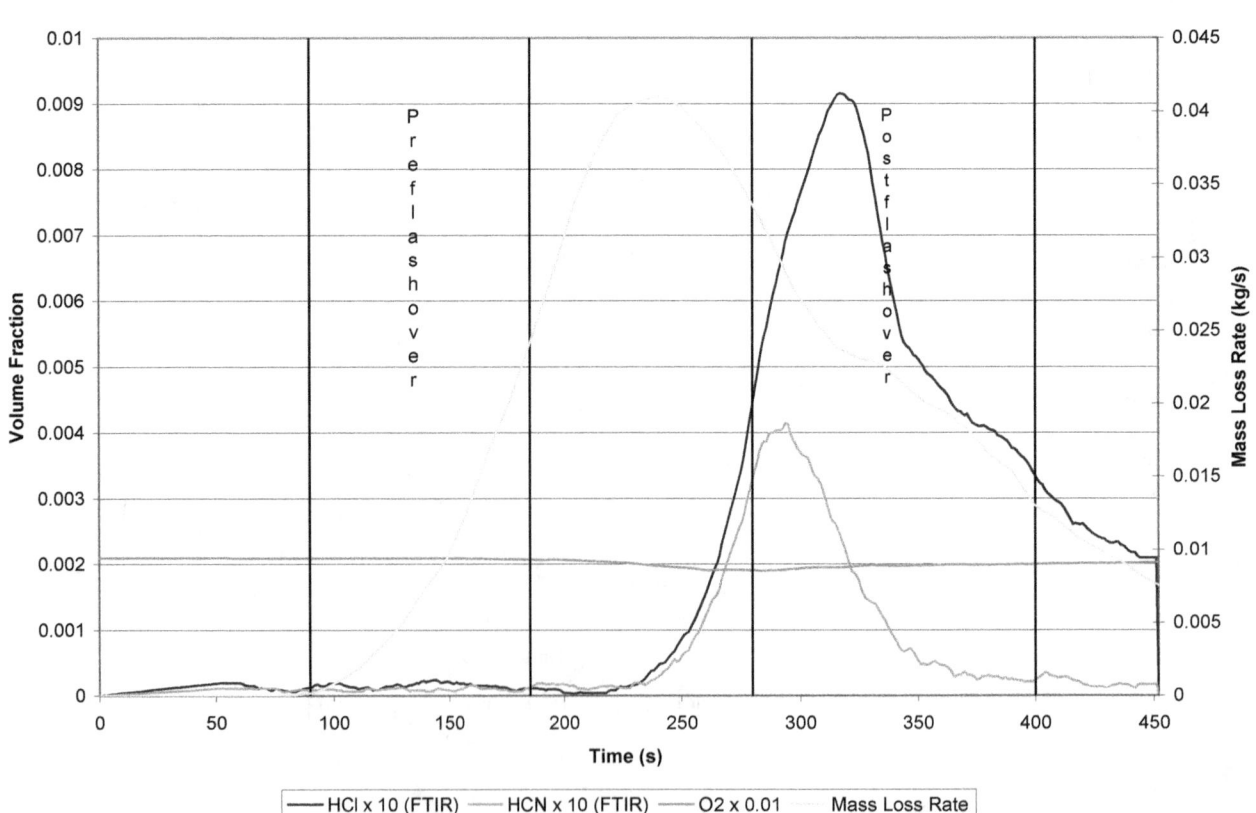

HCl x 10 (FTIR) —— HCN x 10 (FTIR) —— O2 x 0.01 —— Mass Loss Rate

Figures A3c, A3d. Data from Test SW3, Location 4

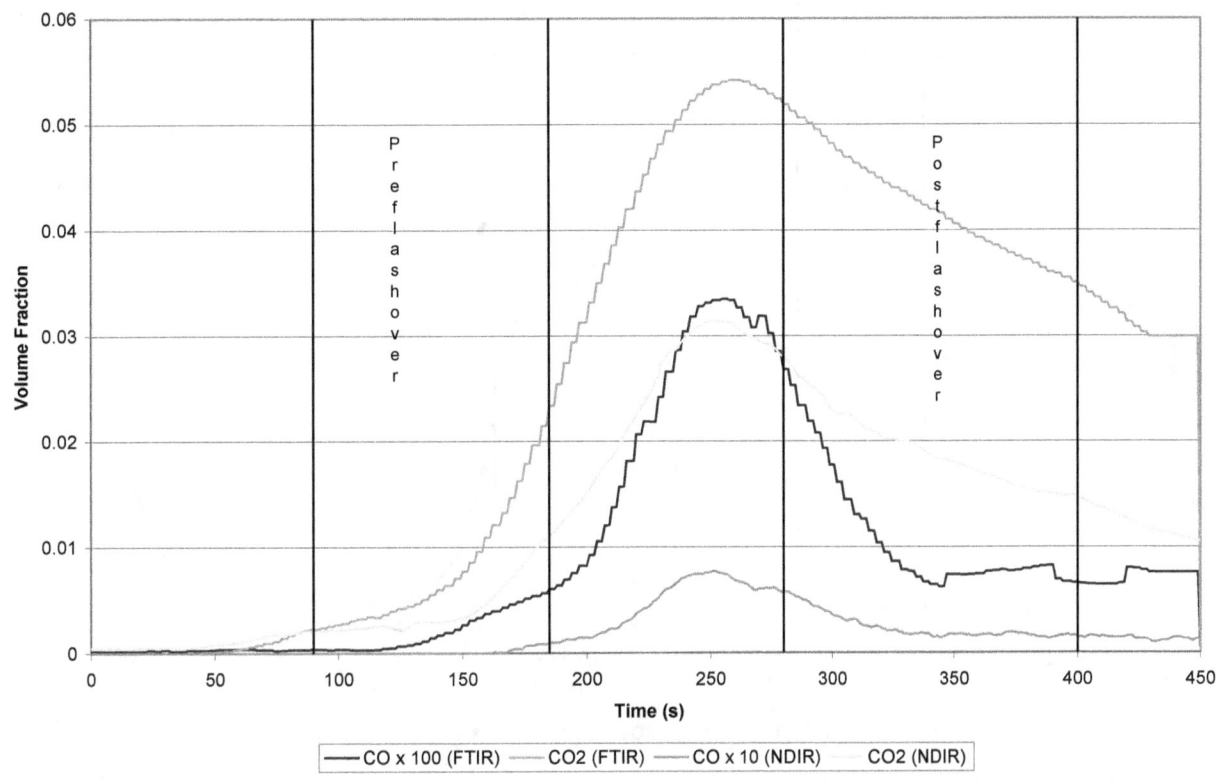

| —— CO x 100 (FTIR) | —— CO2 (FTIR) | —— CO x 10 (NDIR) | CO2 (NDIR) |

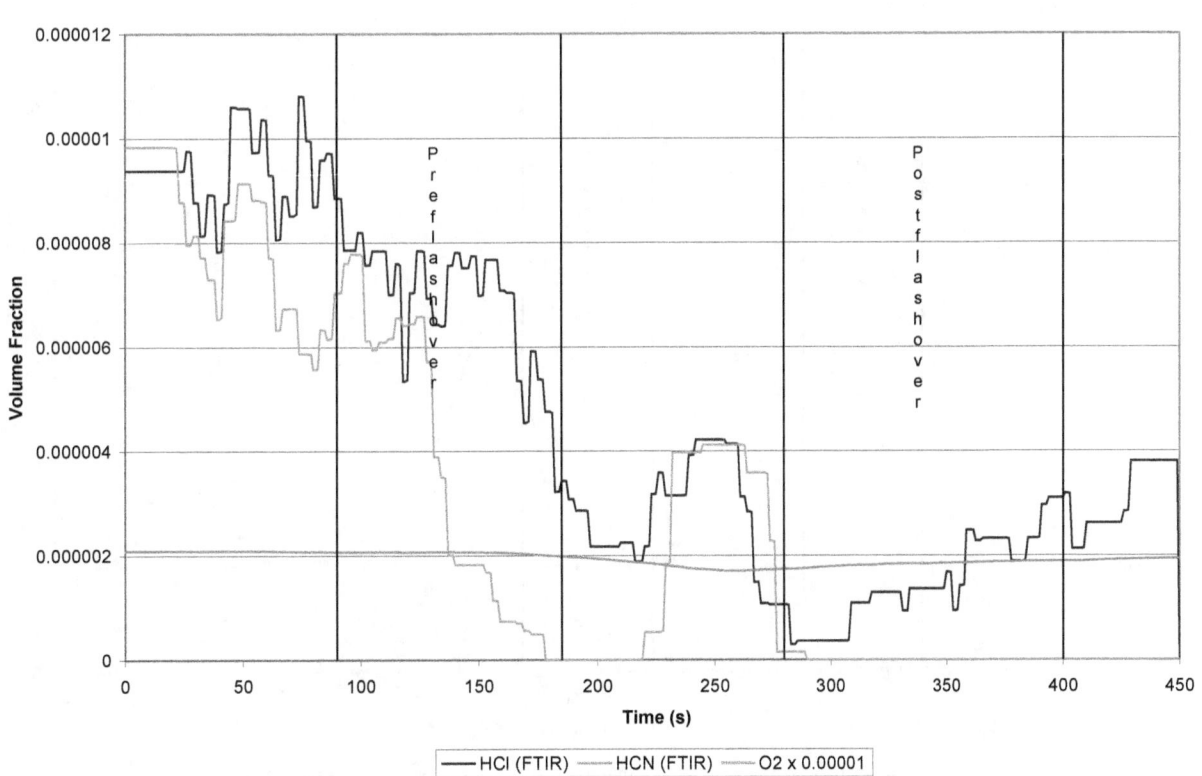

| —— HCl (FTIR) | —— HCN (FTIR) | —— O2 x 0.00001 |

Figures A4a, A4b. Data from Test SW10, Location 2

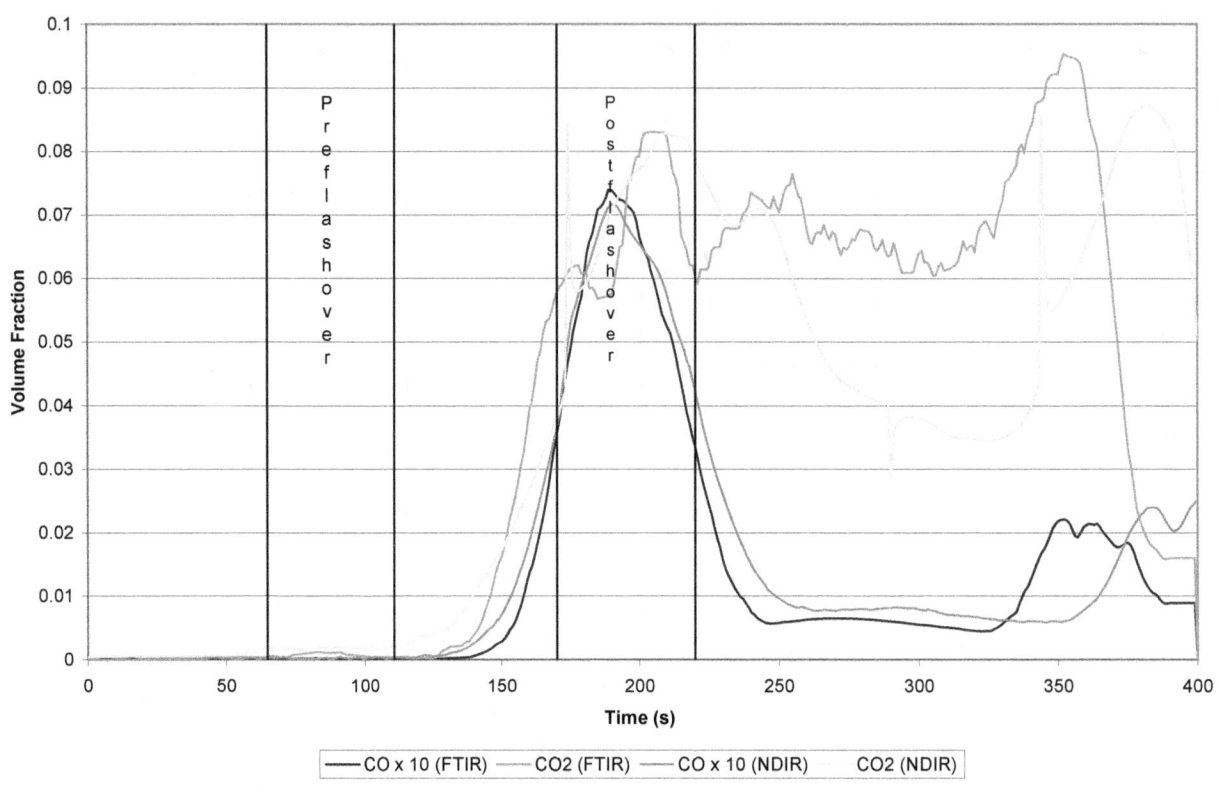

CO x 10 (FTIR) — CO2 (FTIR) — CO x 10 (NDIR) — CO2 (NDIR)

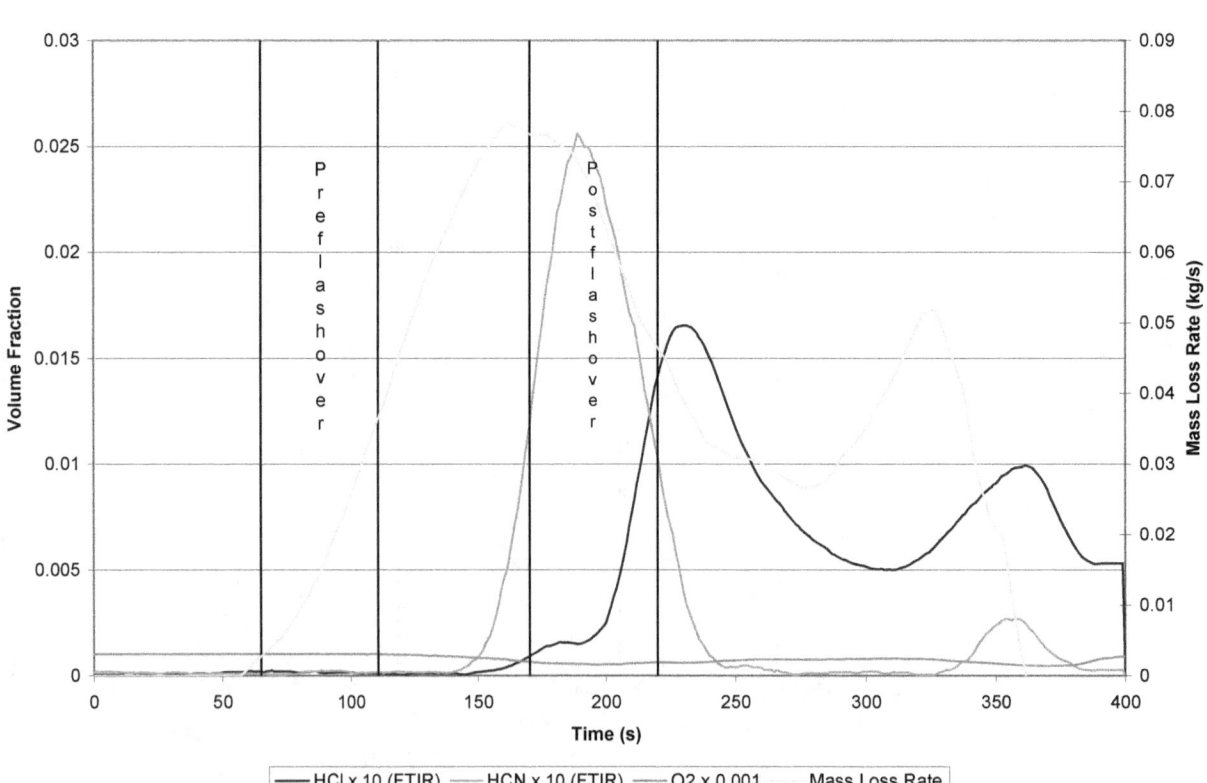

HCl x 10 (FTIR) — HCN x 10 (FTIR) — O2 x 0.001 — Mass Loss Rate

Figures A4c, A4d. Data from Test SW10, Location 4

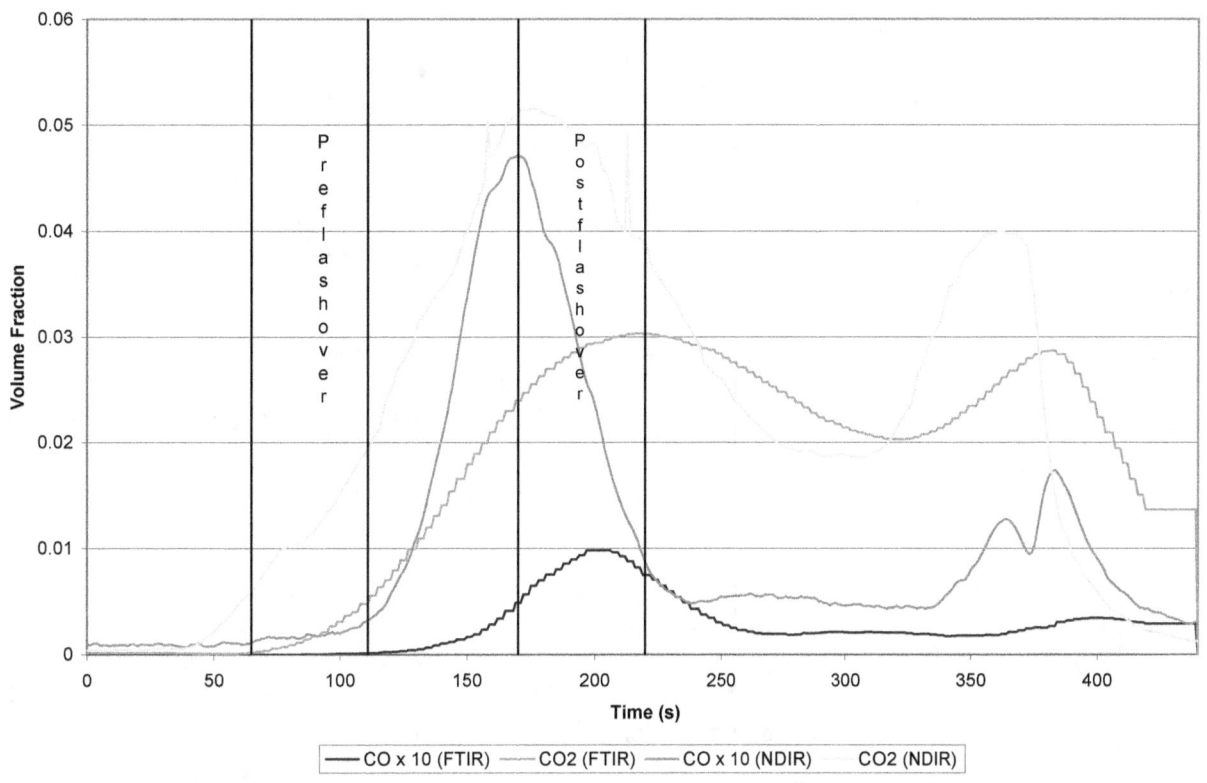

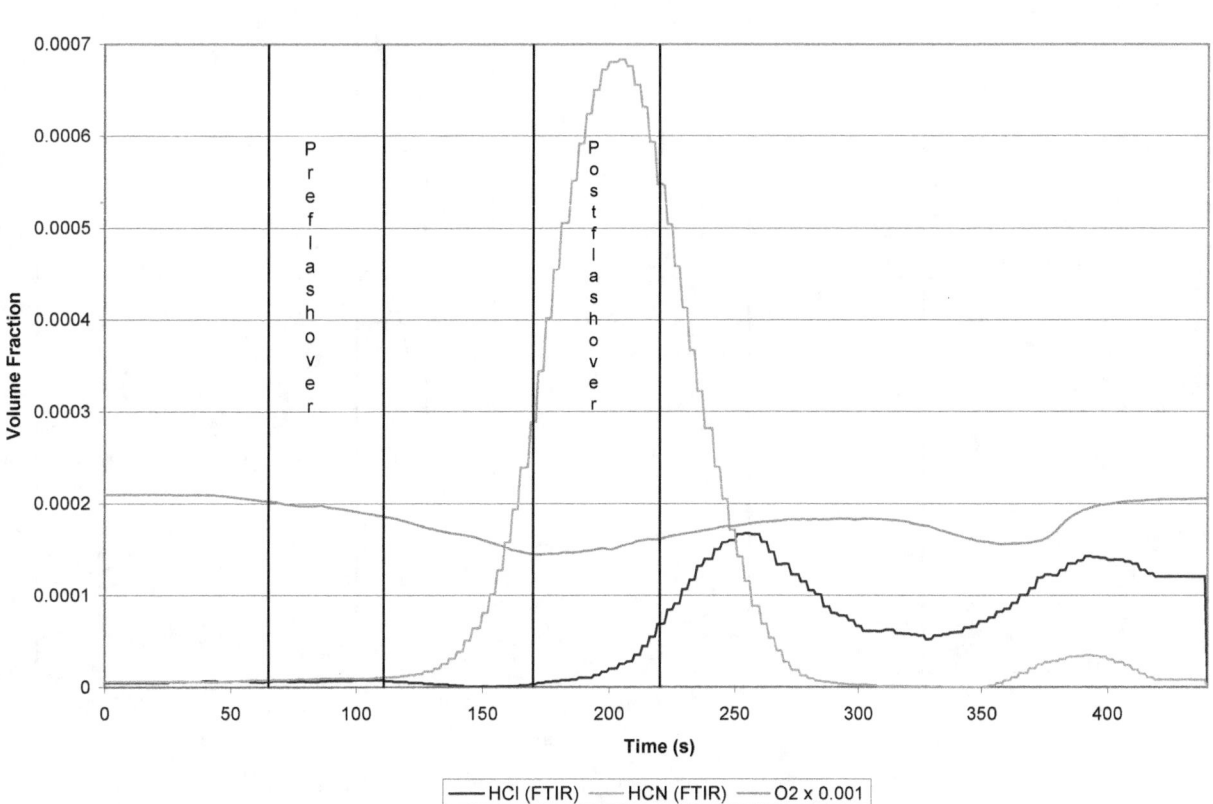

Figures A5a, A5b. Data from Test SW11, Location 2

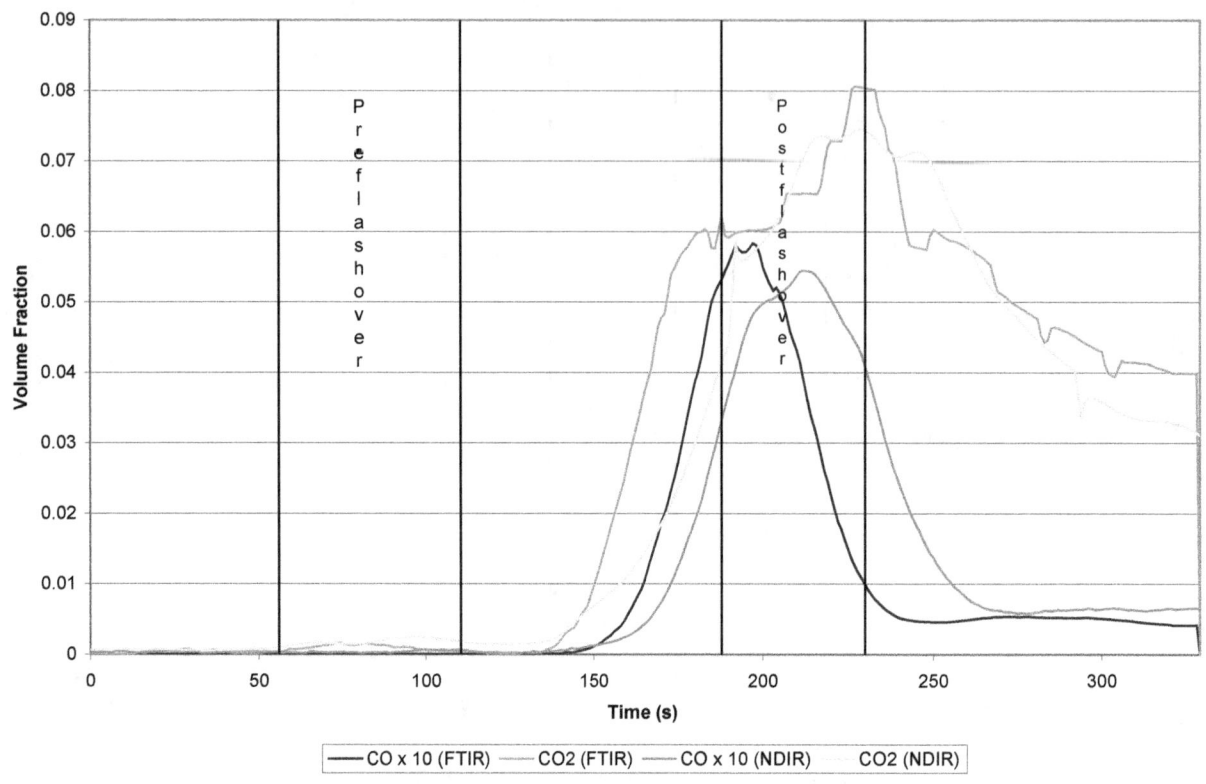

CO x 10 (FTIR) — CO2 (FTIR) — CO x 10 (NDIR) — CO2 (NDIR)

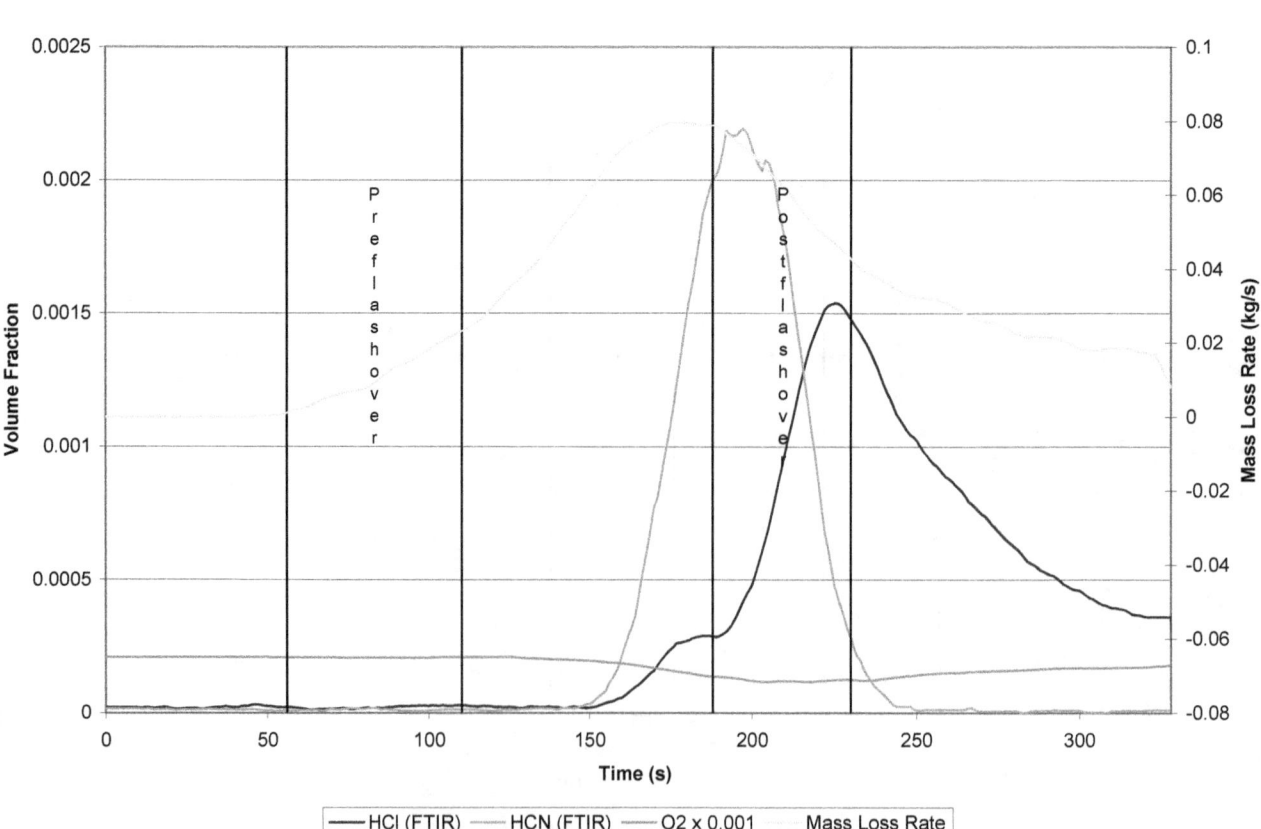

HCl (FTIR) — HCN (FTIR) — O2 x 0.001 — Mass Loss Rate

Figures A5c, A5d. Data from Test SW11, Location 4

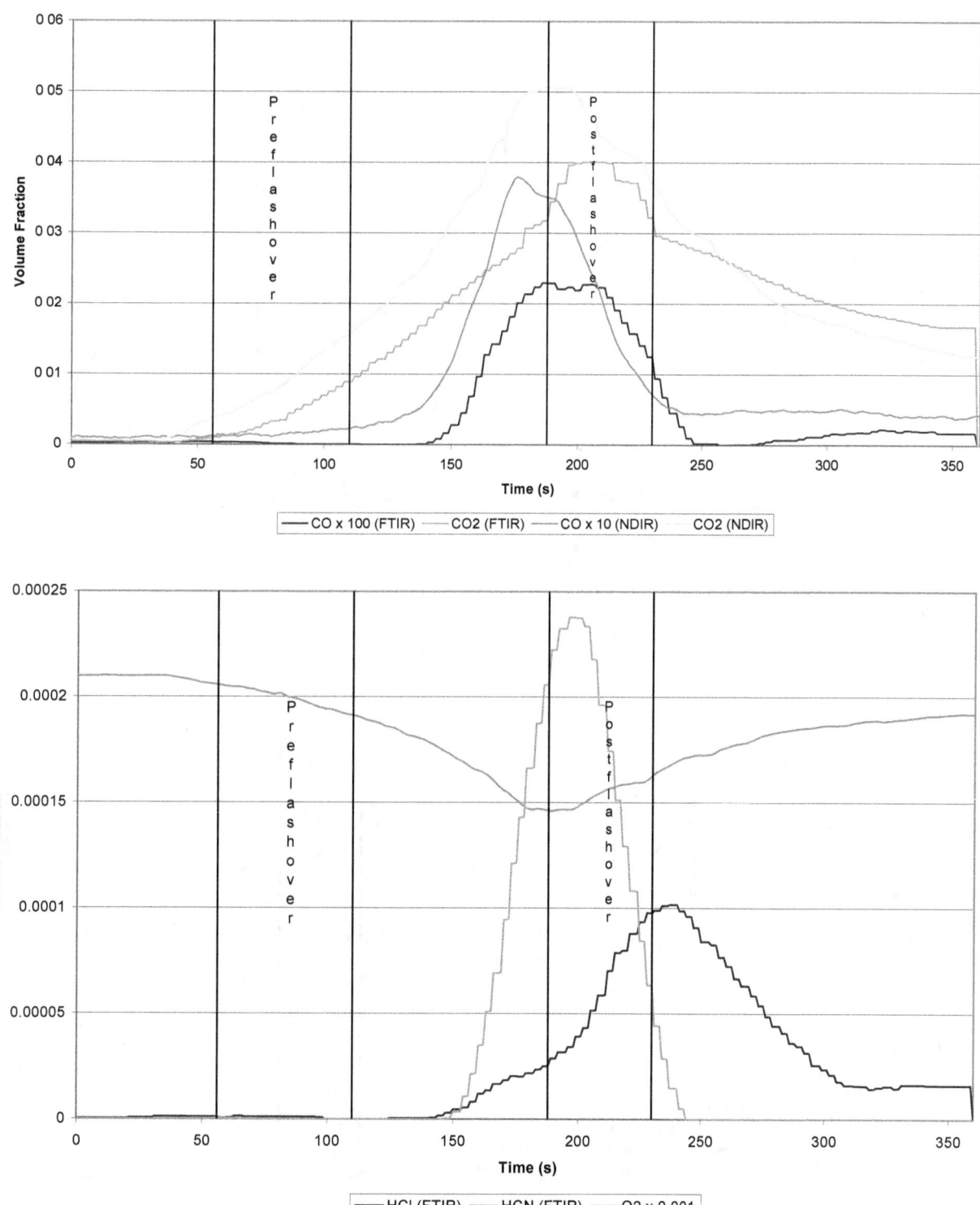

Figures A6a, A6b. Data from Test SW12, Location 2

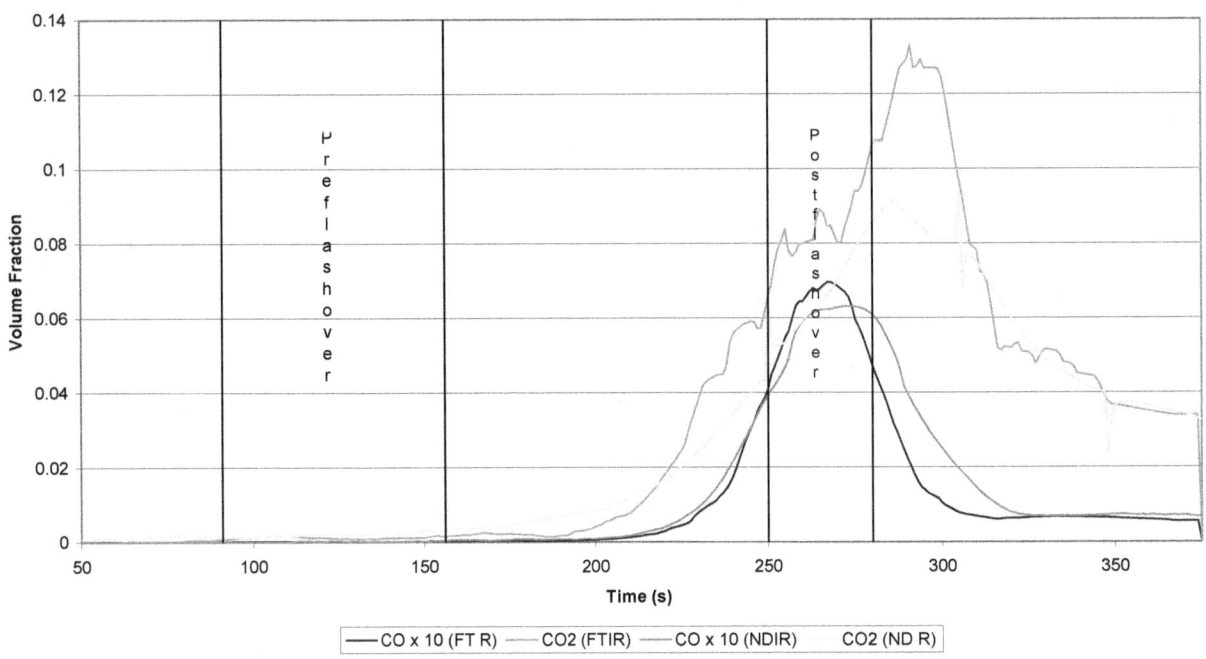

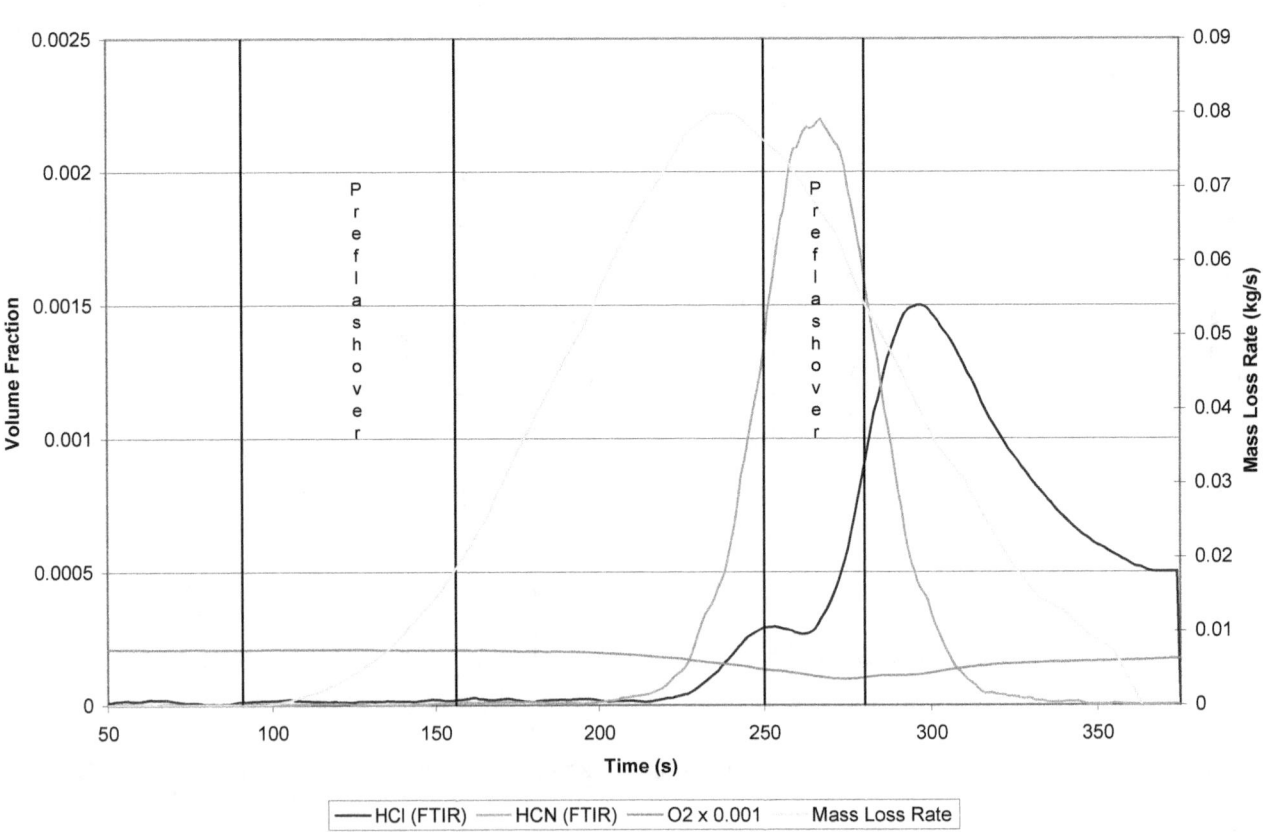

Figures A6c, A6d. Data from Test SW12, Location 4

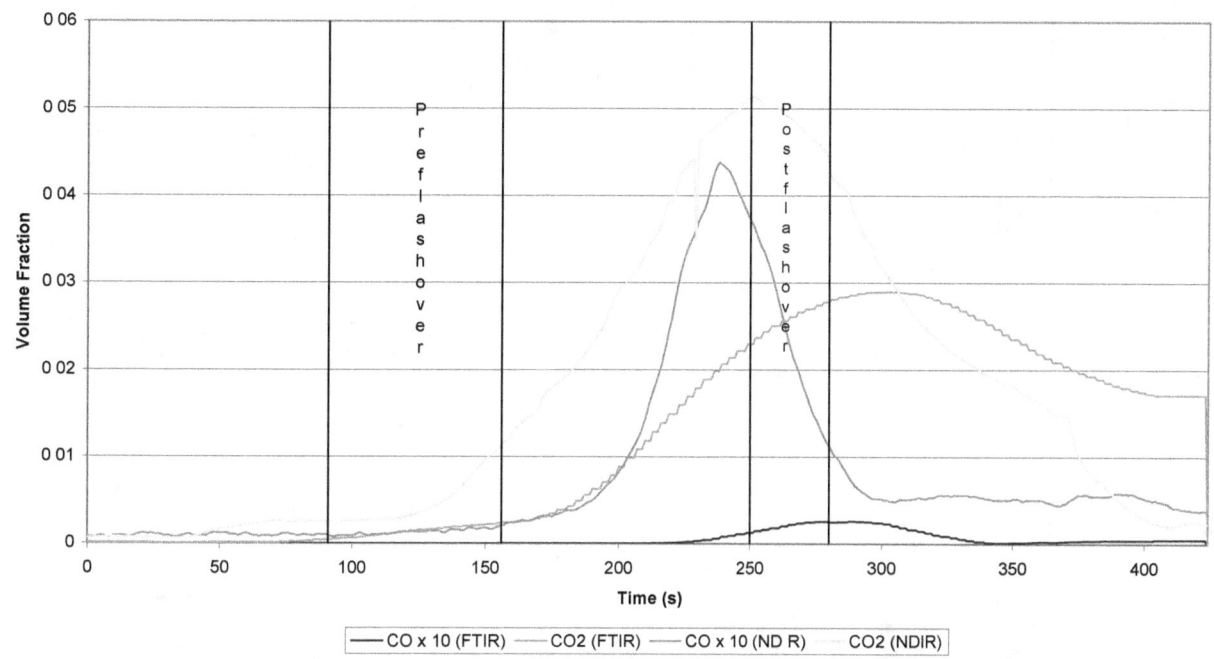

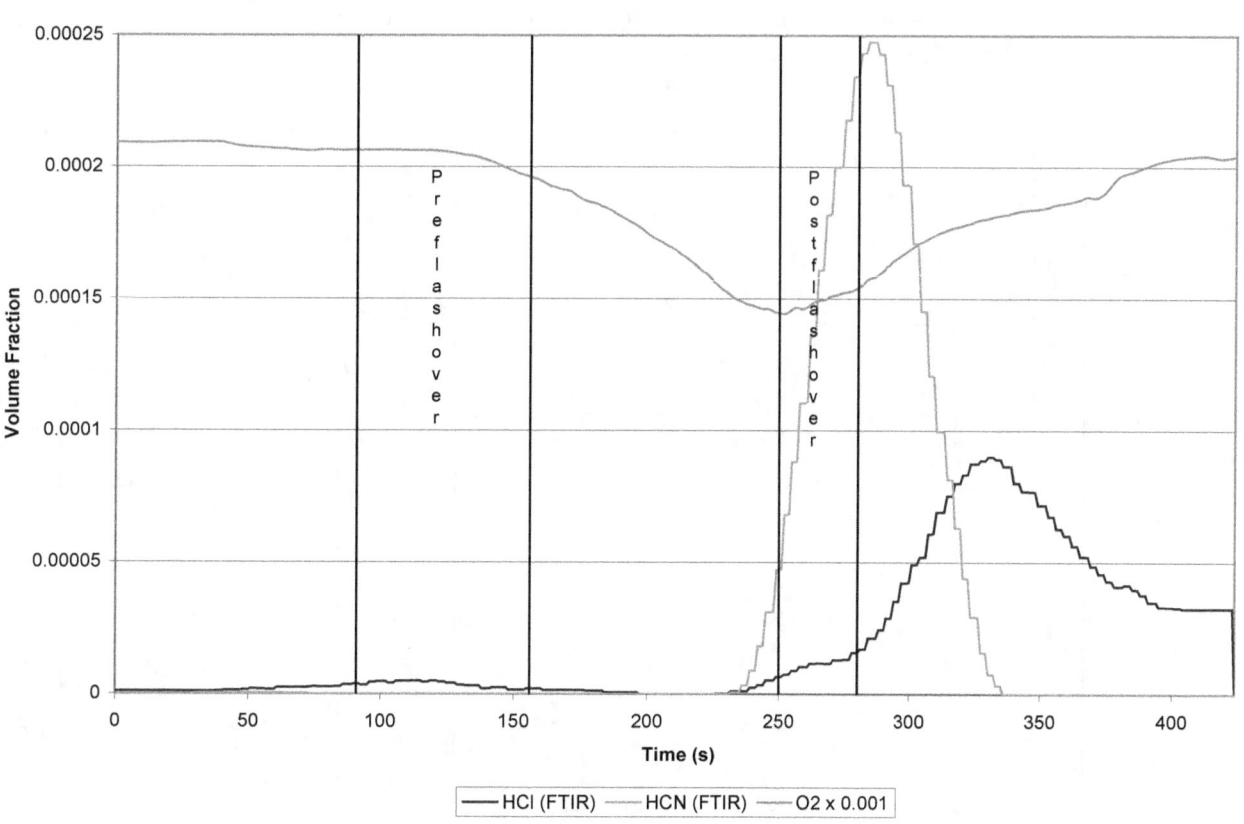

Figures A7a, A7b. Data from Test SW13, Location 2

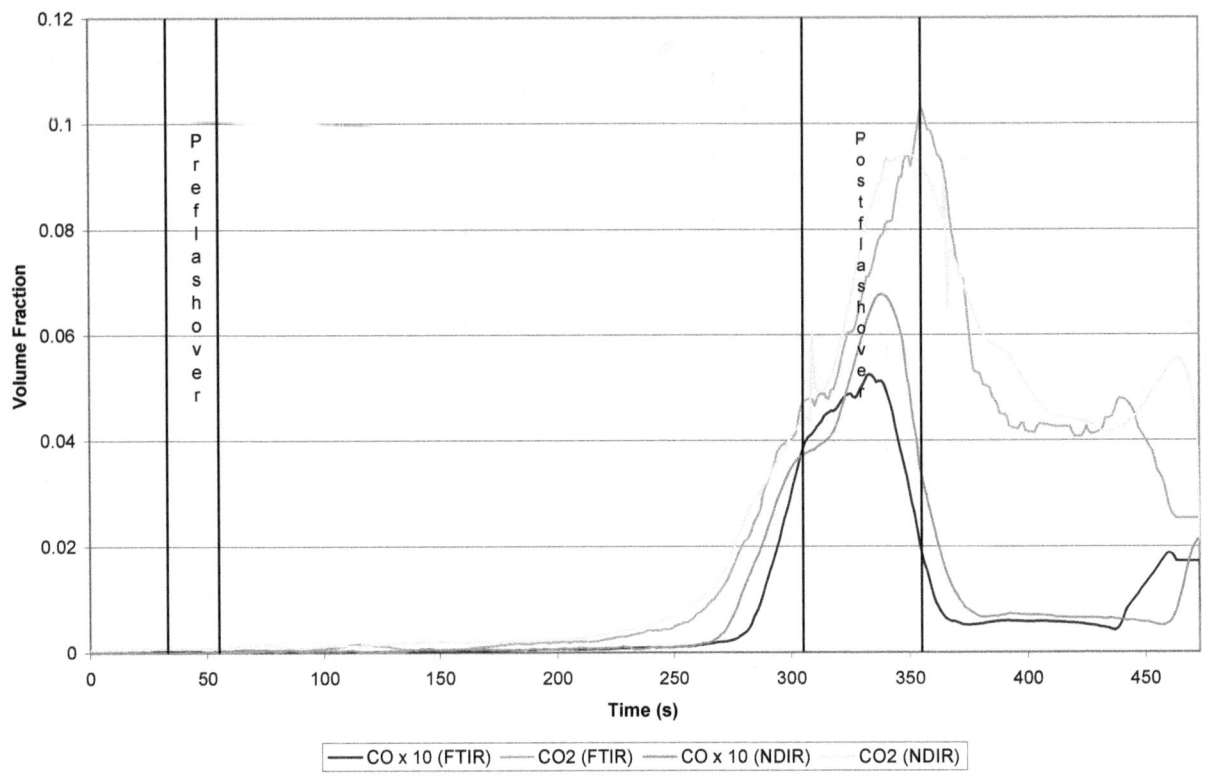

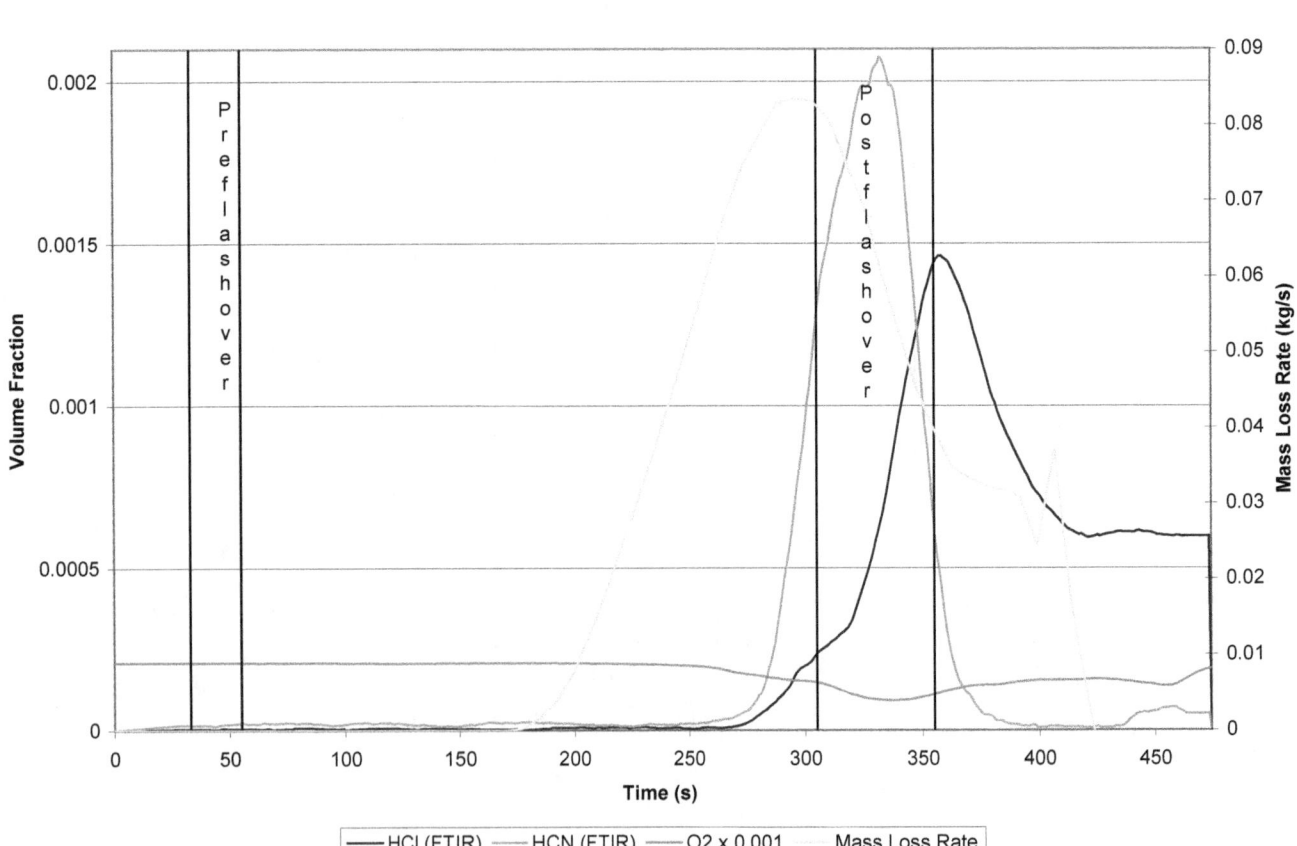

Figures A7c, A7d. Data from Test SW13, Location 4

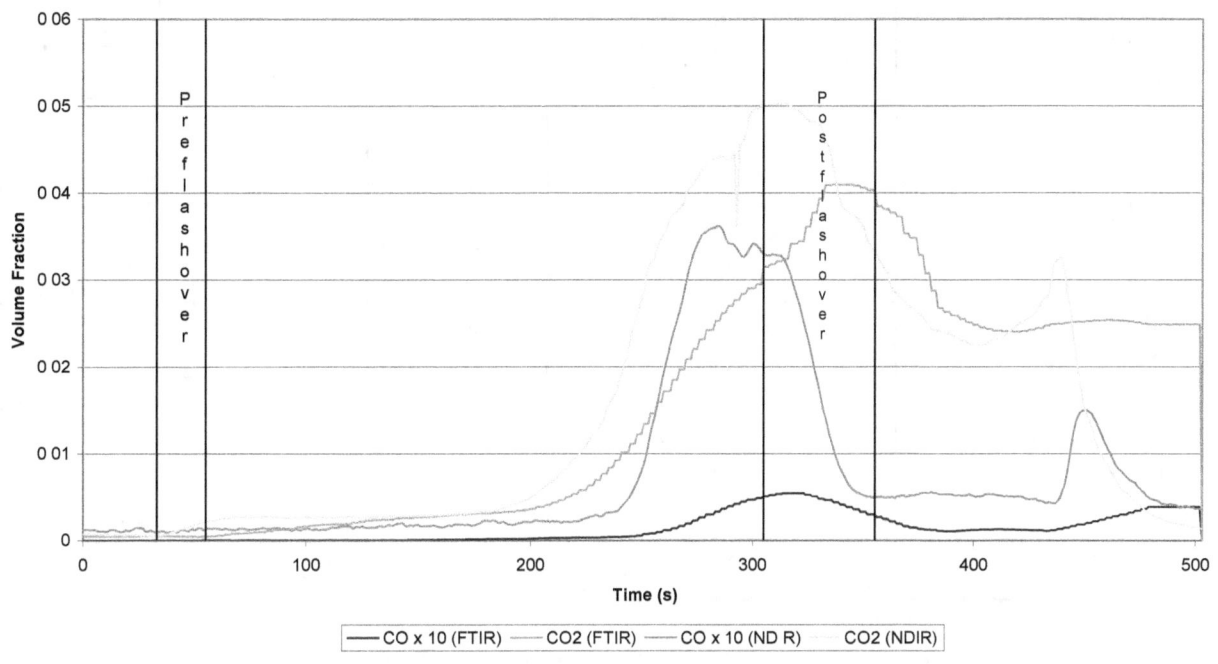

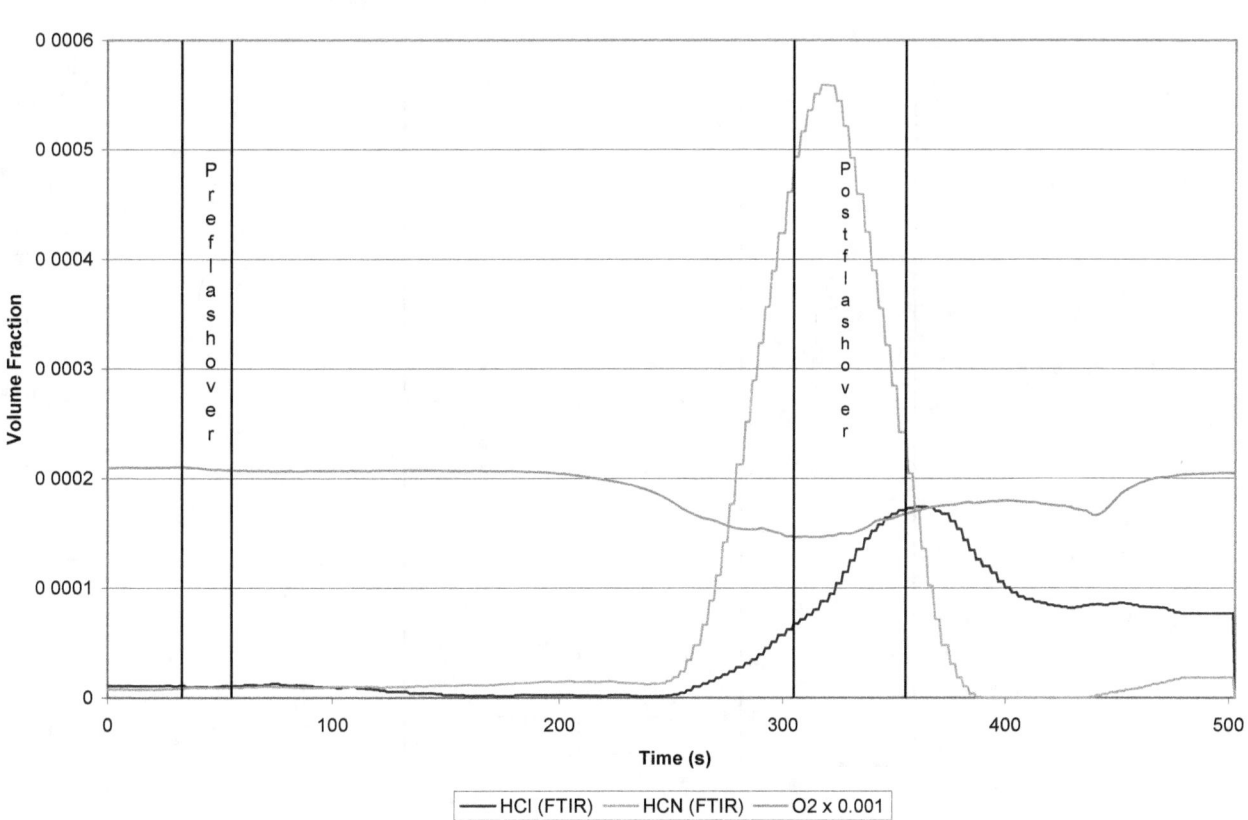

Figures A8a, A8b. Data from Test SW14, Location 2

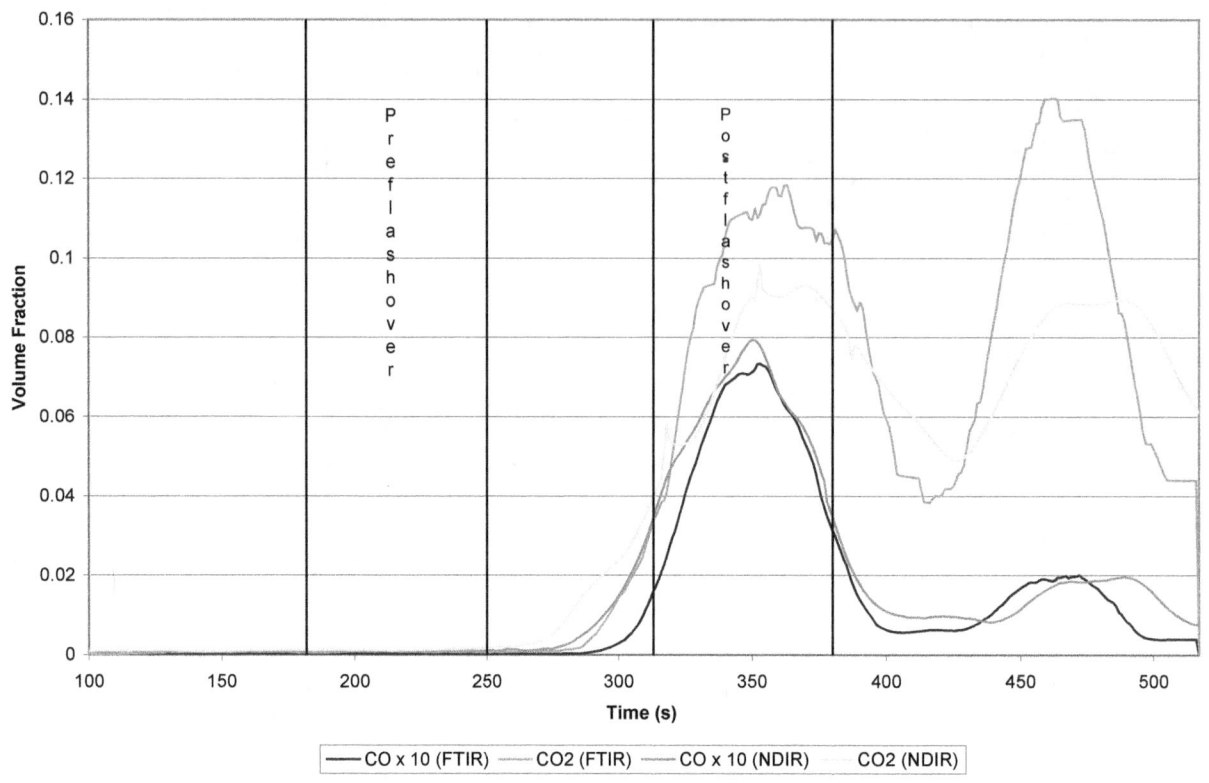

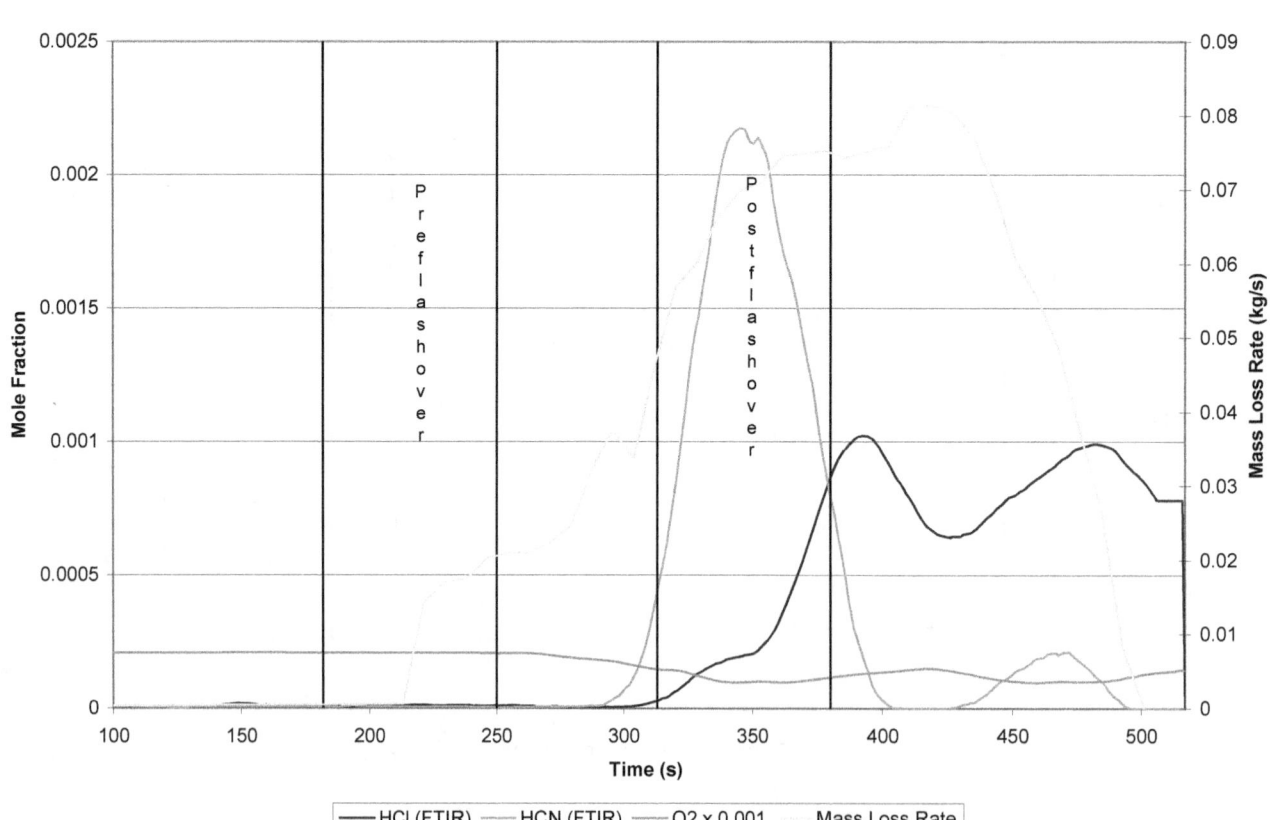

Figures A8c, A8d. Data from Test SW14, Location 4

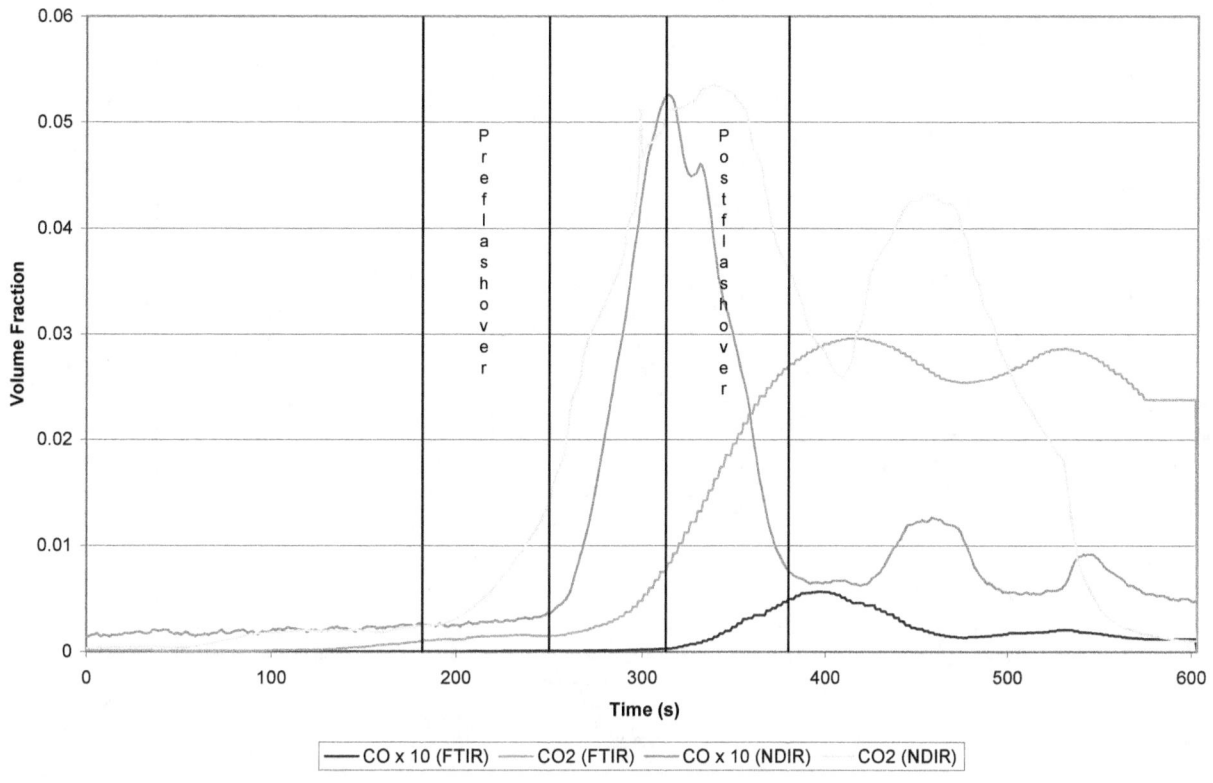

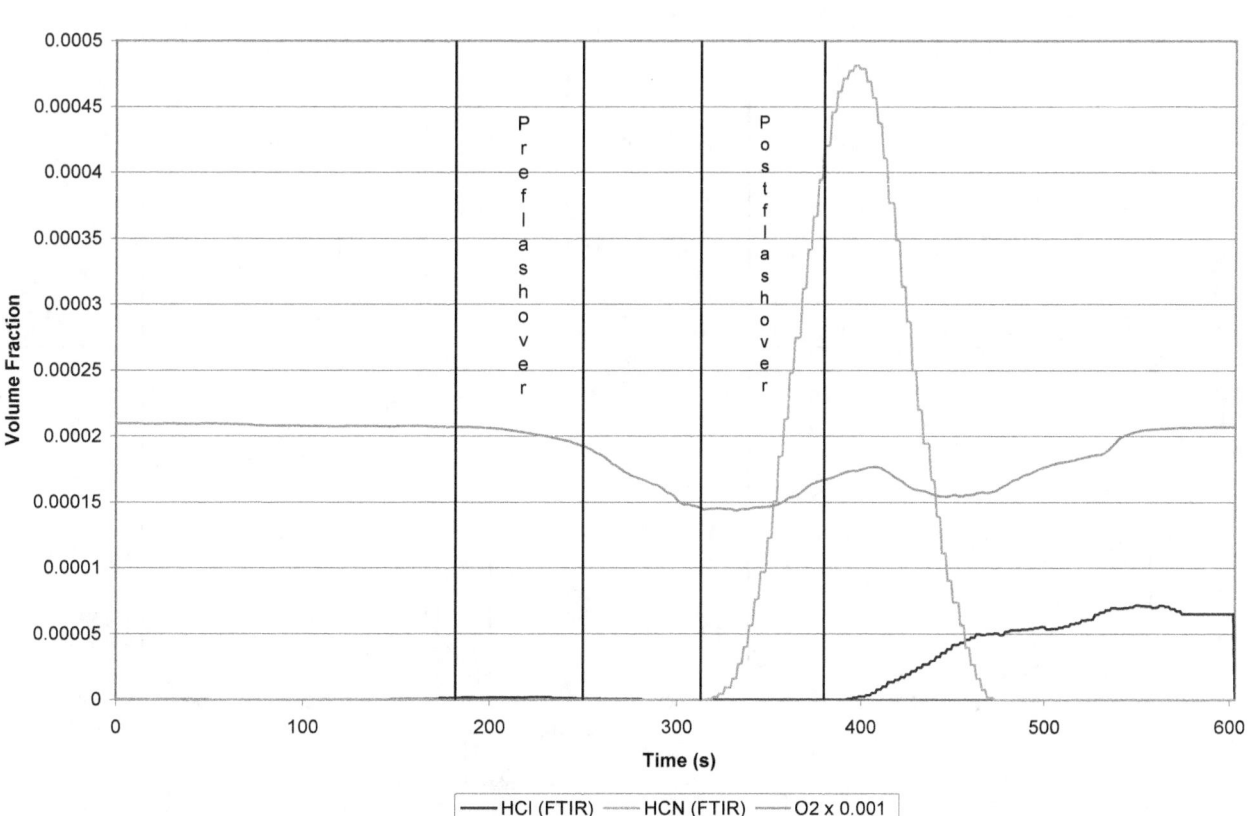

Figures A9a, A9b. Data from Test BW1, Location 2

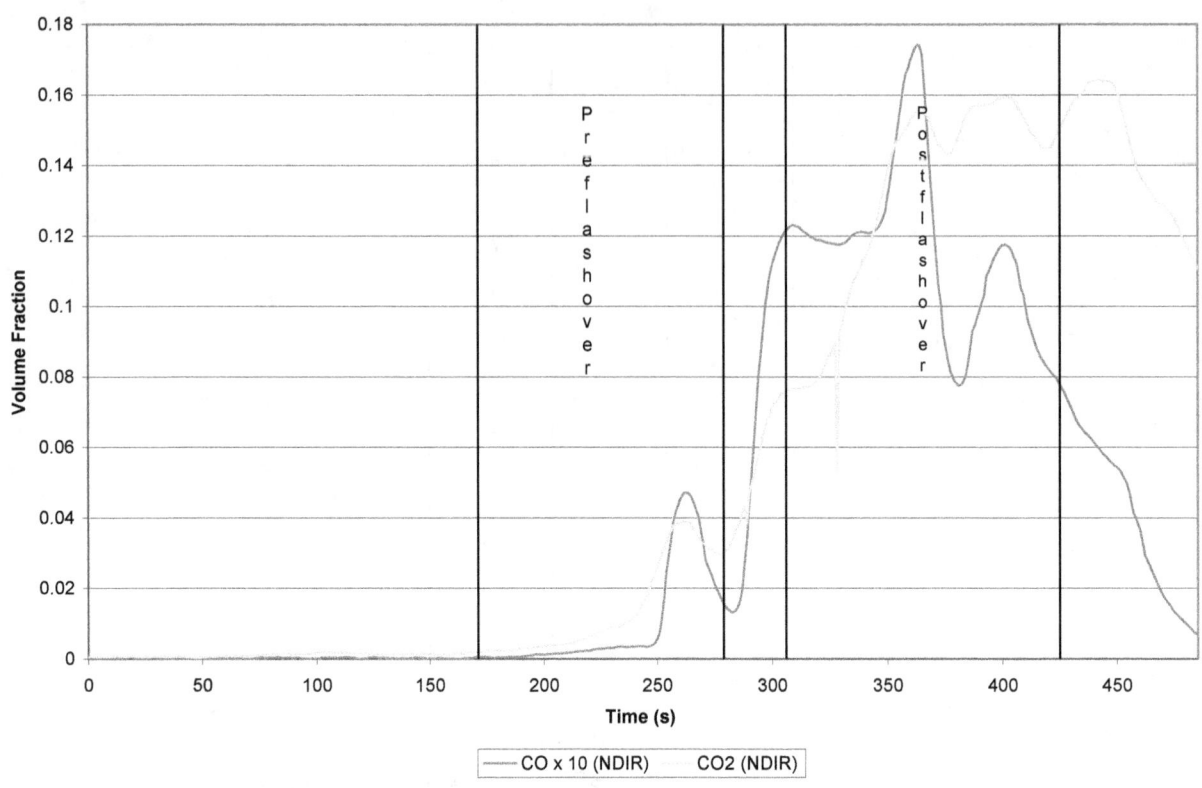

CO x 10 (NDIR)   CO2 (NDIR)

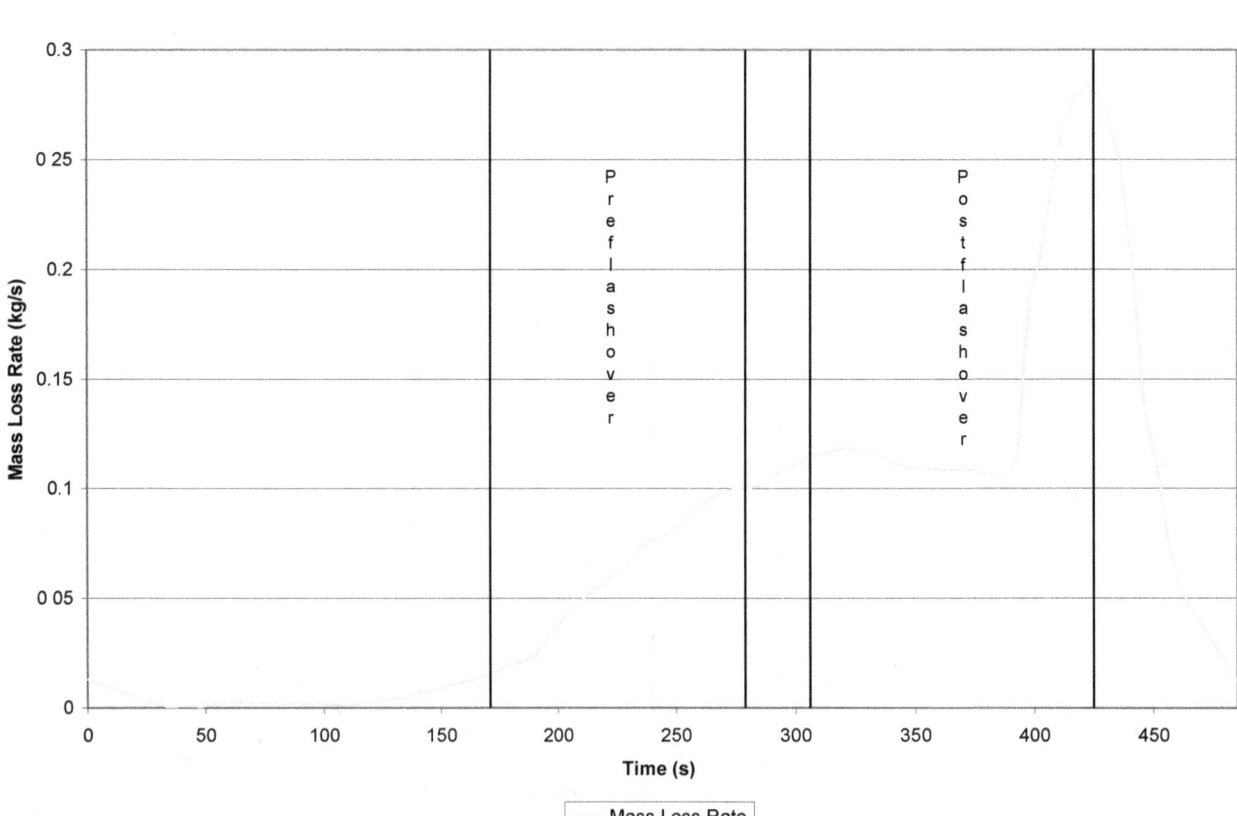

Mass Loss Rate

Figures A10a, A10b. Data from Test BW2, Location 2

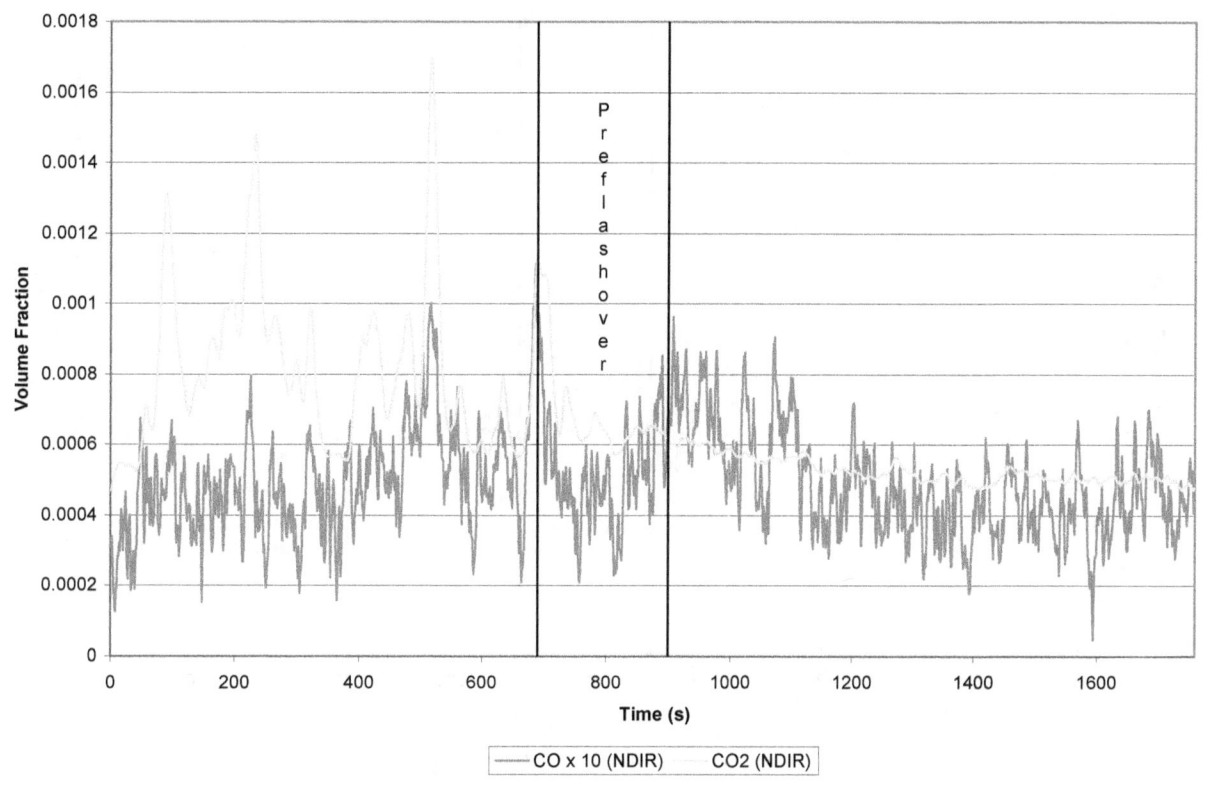

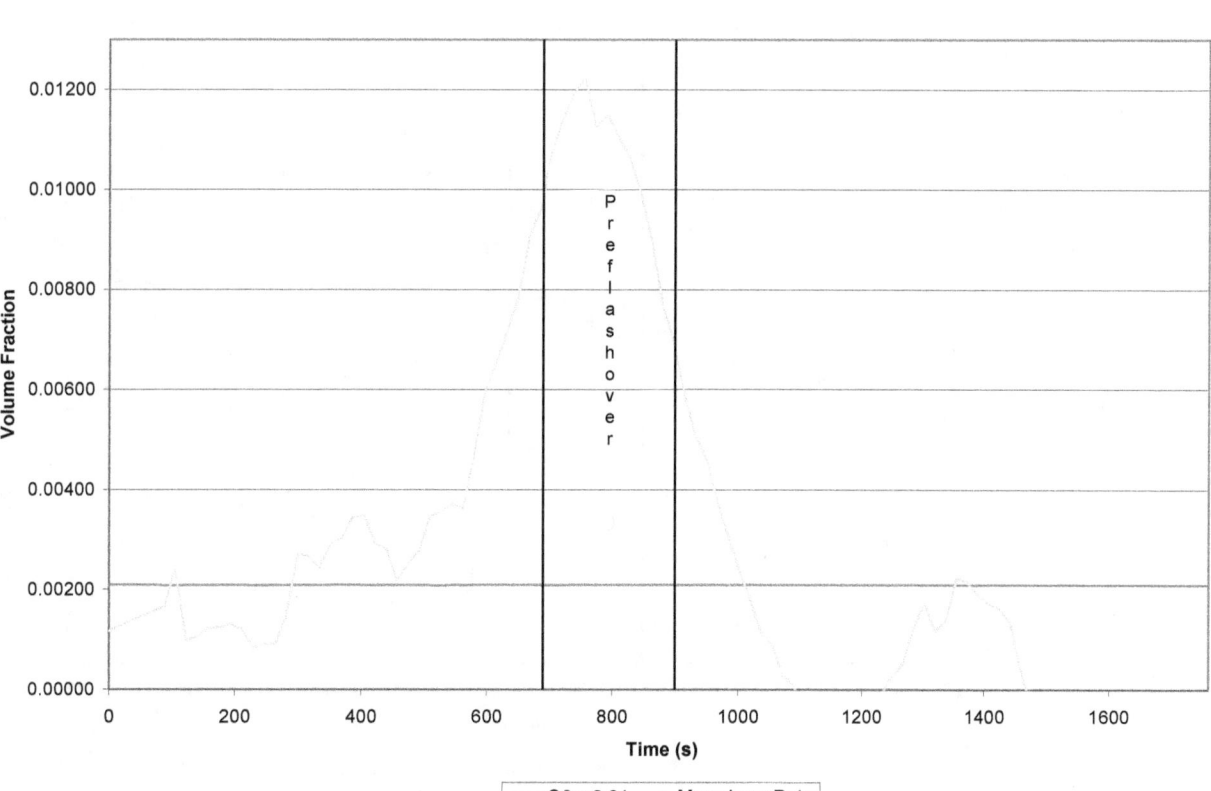

Figures A10c, A10d. Data from Test BW2, Location 4

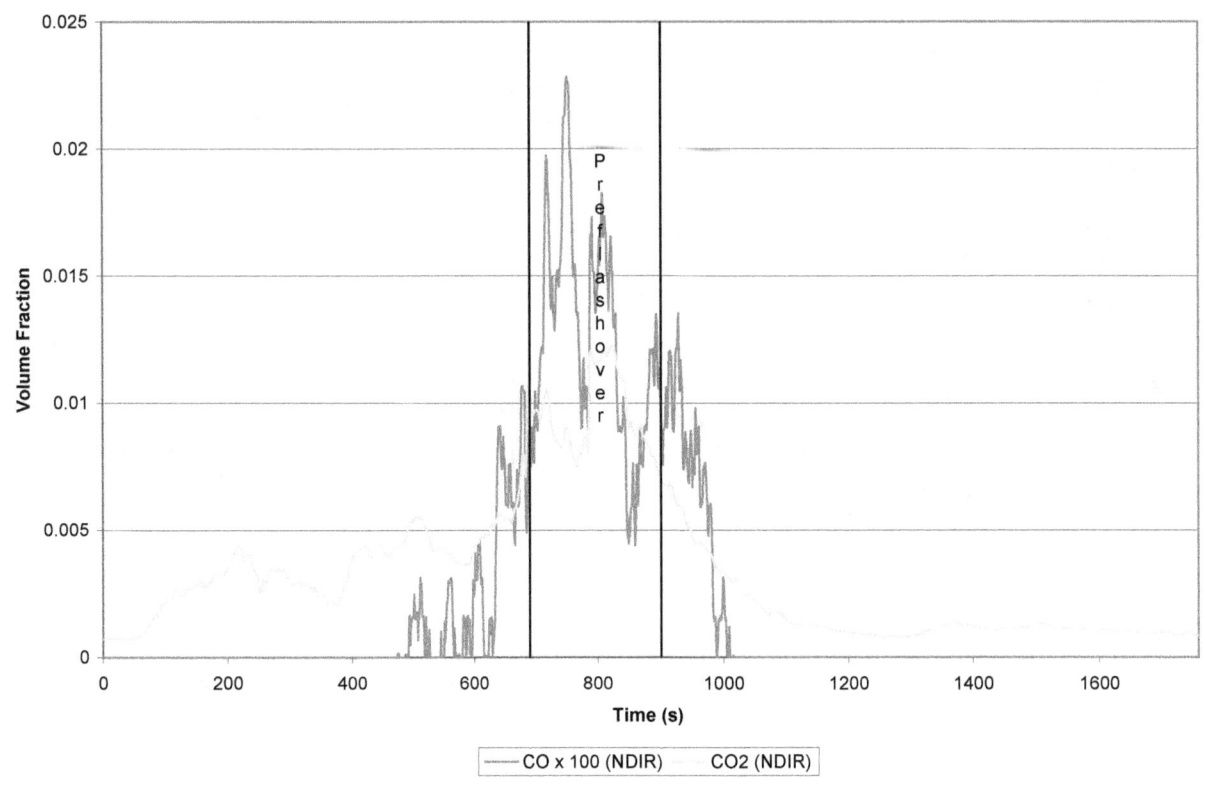

Preflashover

CO x 100 (NDIR)   CO2 (NDIR)

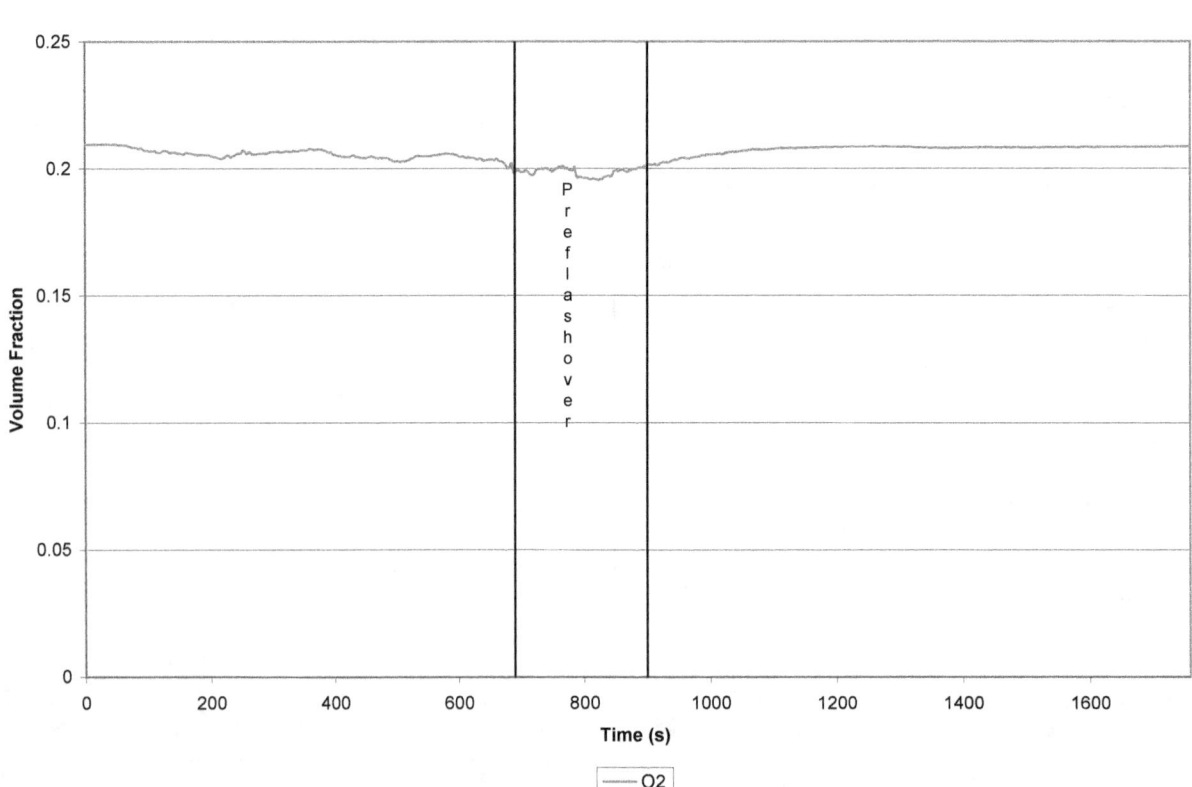

Preflashover

O2

Figures A11a, A11b. Data from Test BW3, Location 2

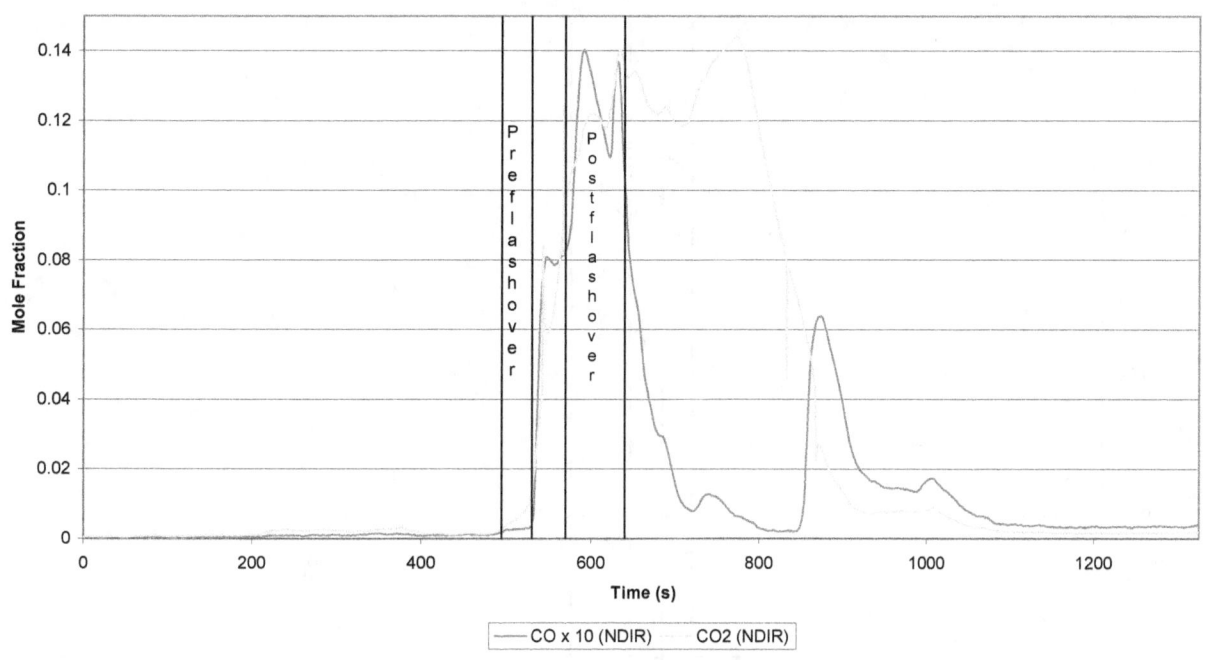

CO x 10 (NDIR)   CO2 (NDIR)

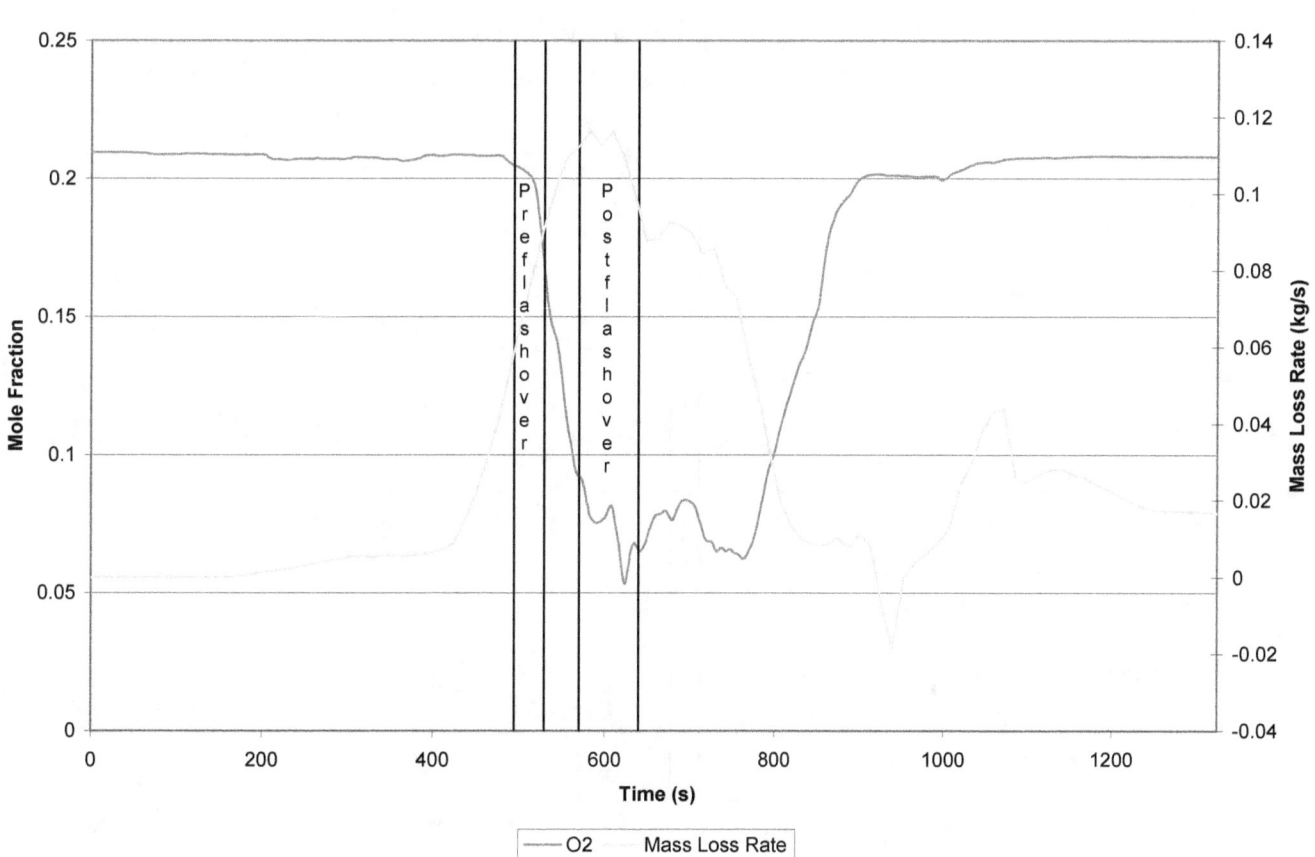

O2   Mass Loss Rate

Figures A11c, A11d. Data from Test BW3, Location 4

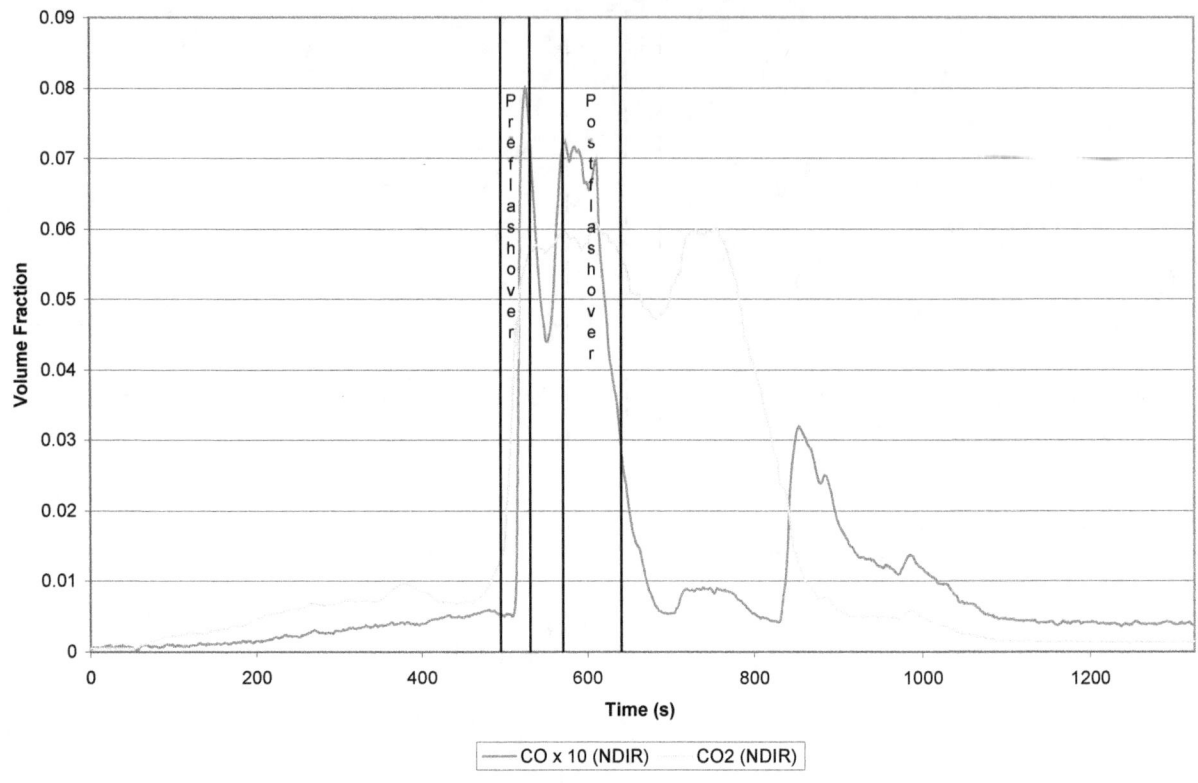

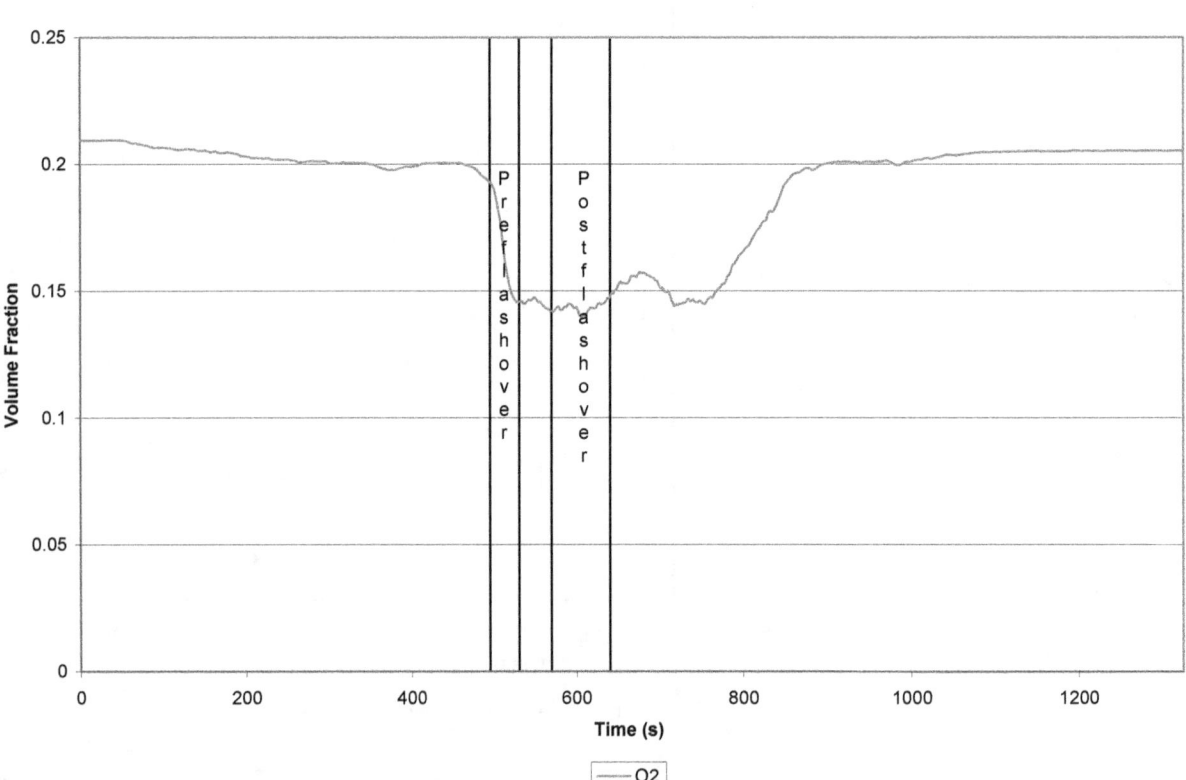

Figures A12a, A12b. Data from Test BW4, Location 2

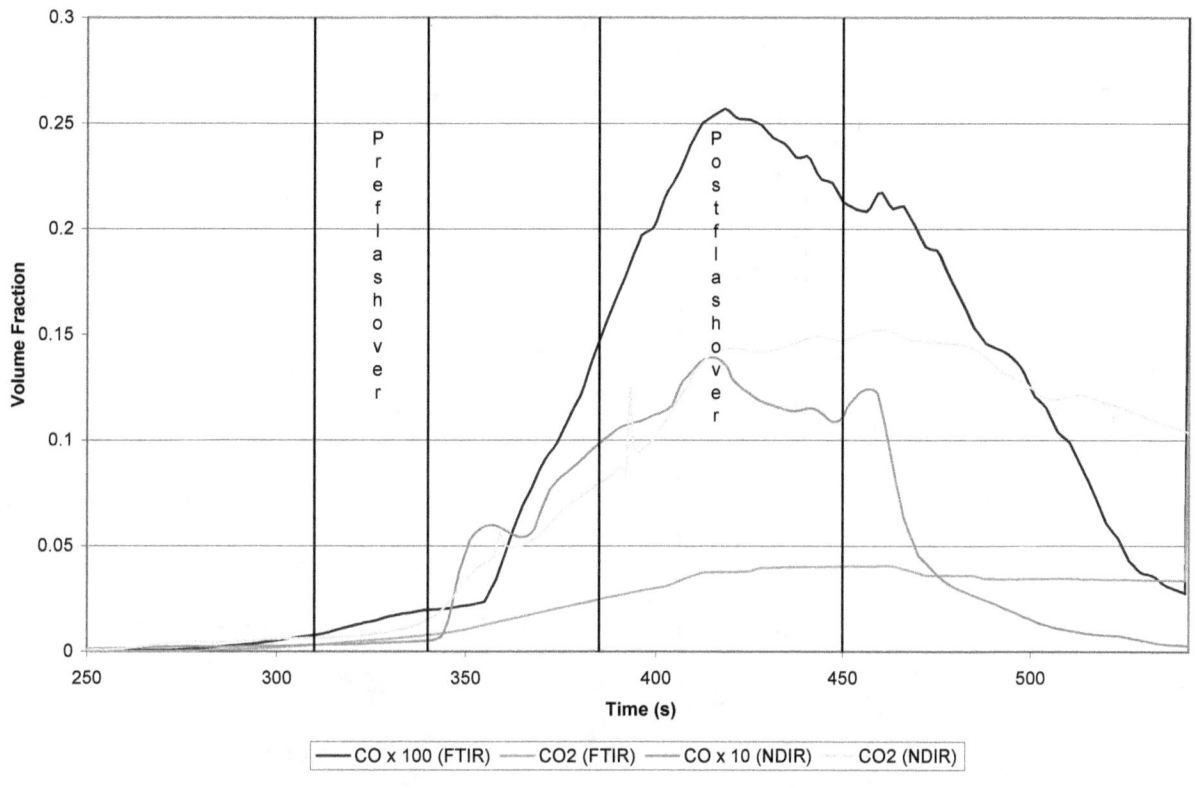

Legend: CO x 100 (FTIR) — CO2 (FTIR) — CO x 10 (NDIR) — CO2 (NDIR)

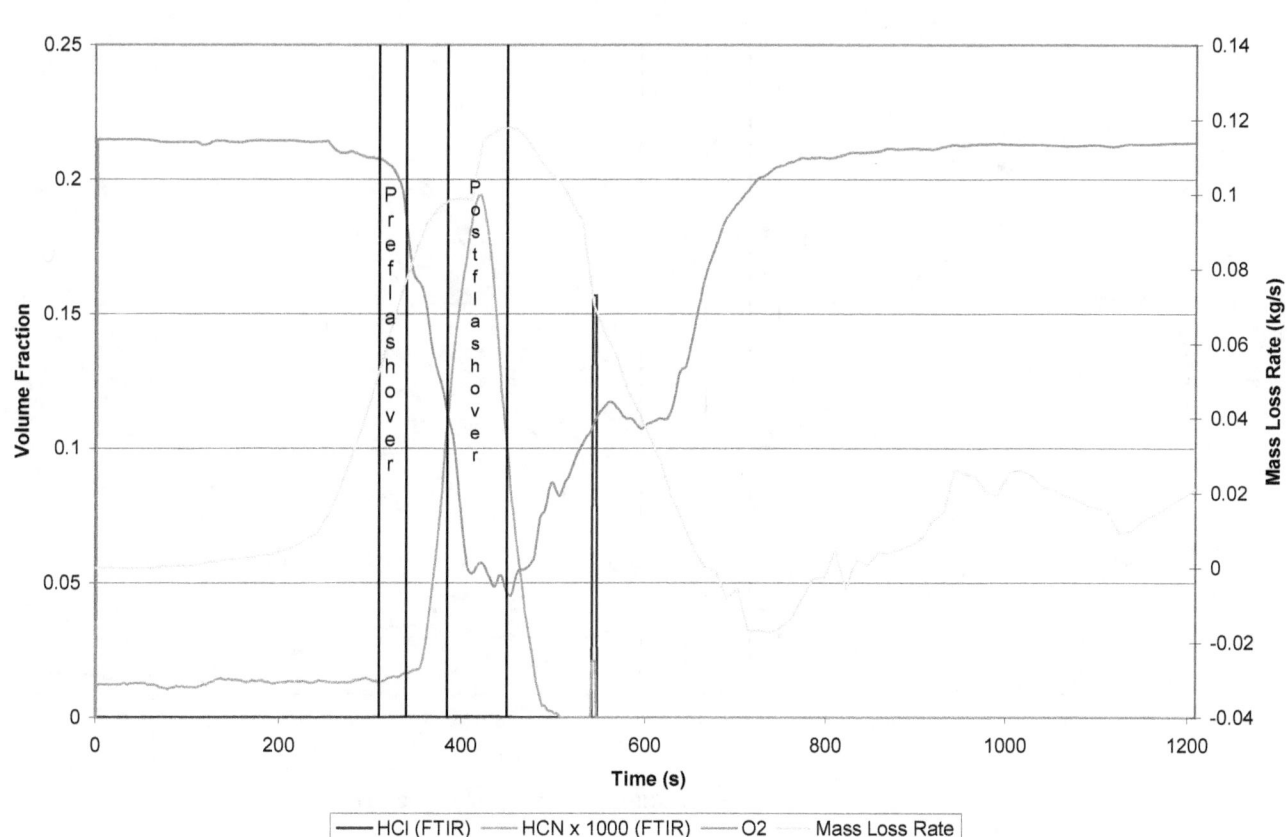

Legend: HCl (FTIR) — HCN x 1000 (FTIR) — O2 — Mass Loss Rate

Figures A12c, A12d. Data from Test BW4, Location 4

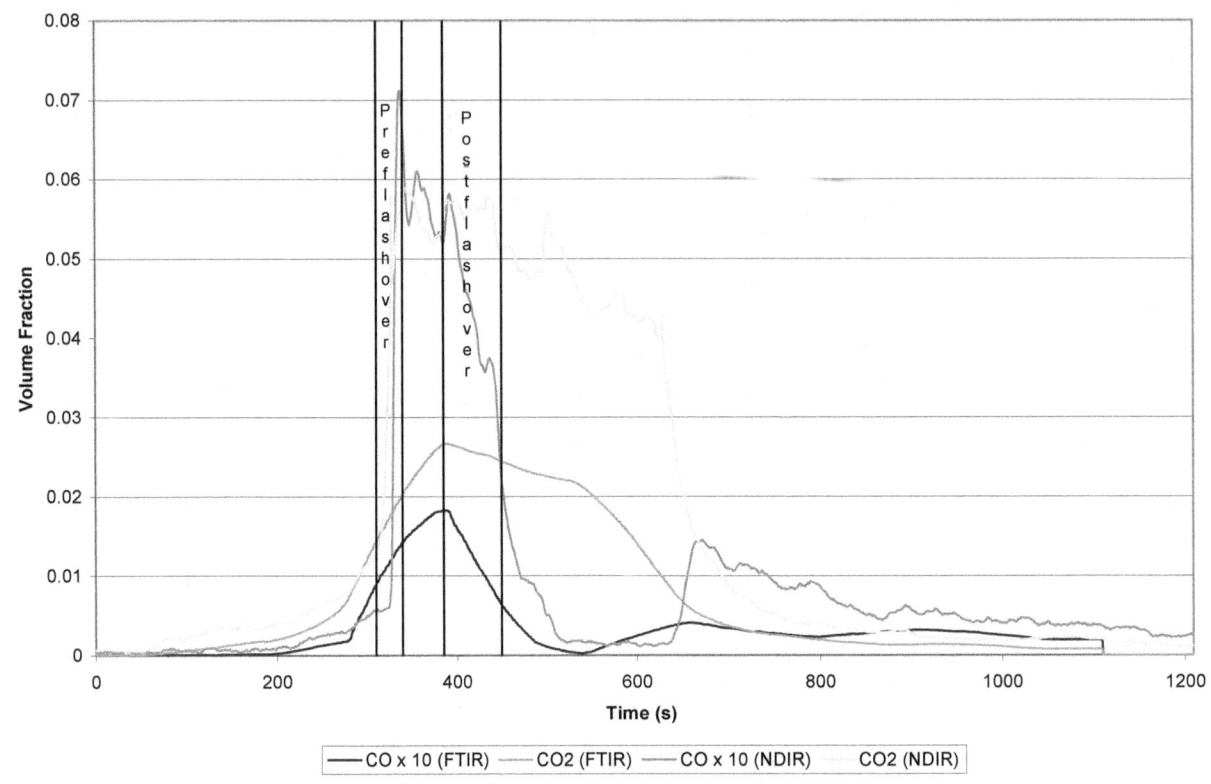

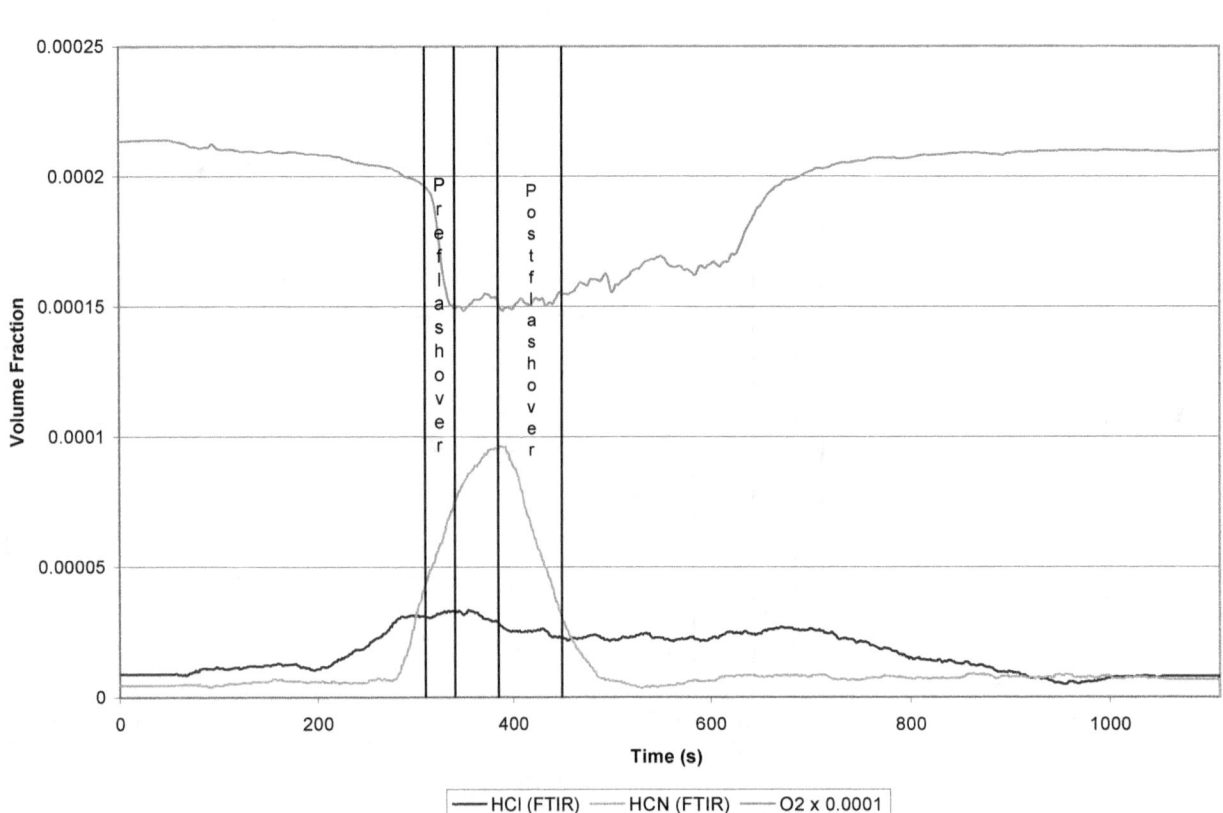

Figures A13a, A13b. Data from Test BW5, Location 2

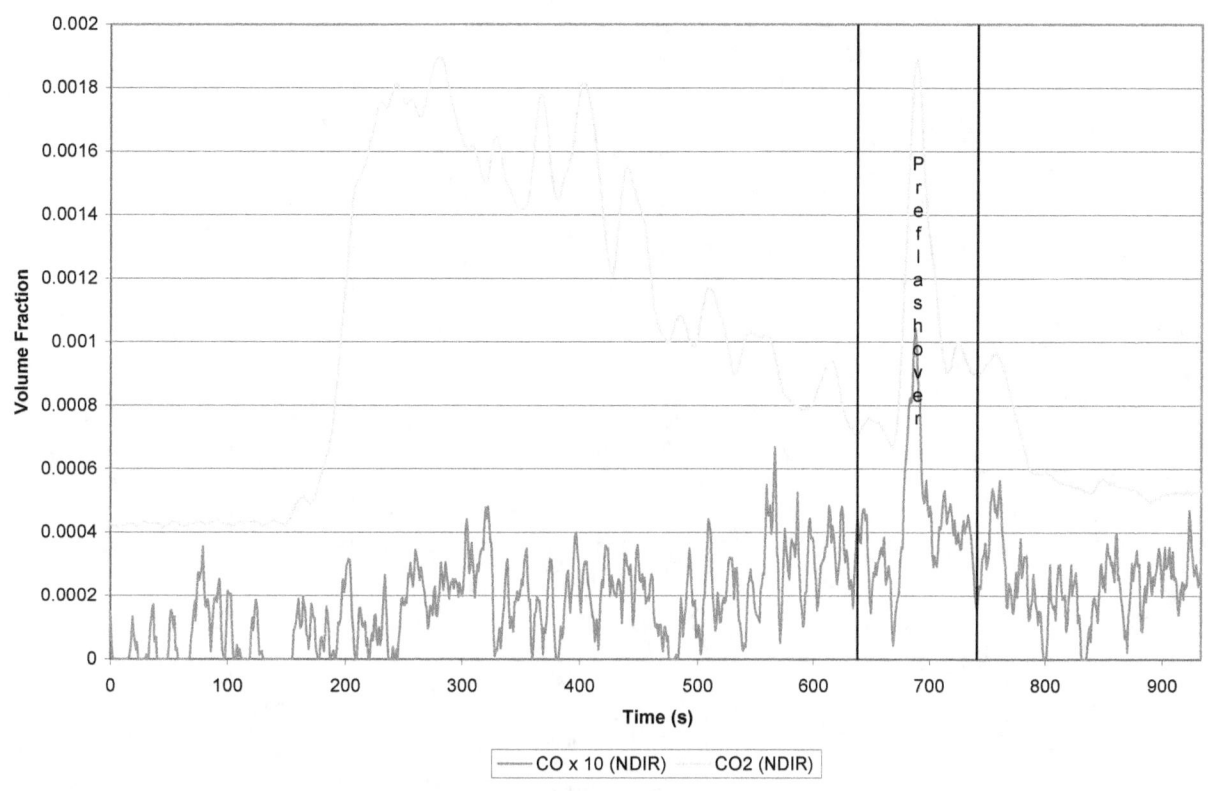

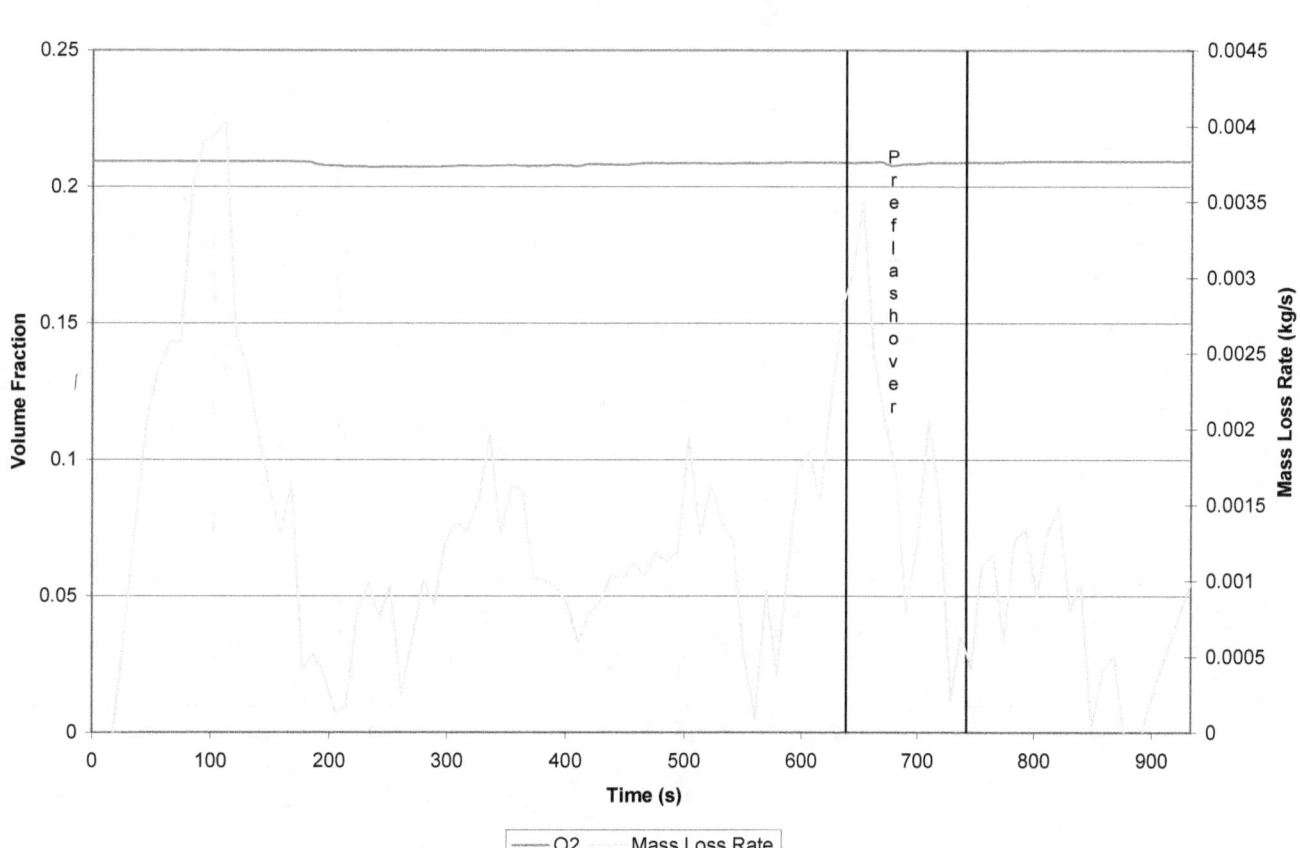

Figures A13c, A13d. Data from Test BW5, Location 4

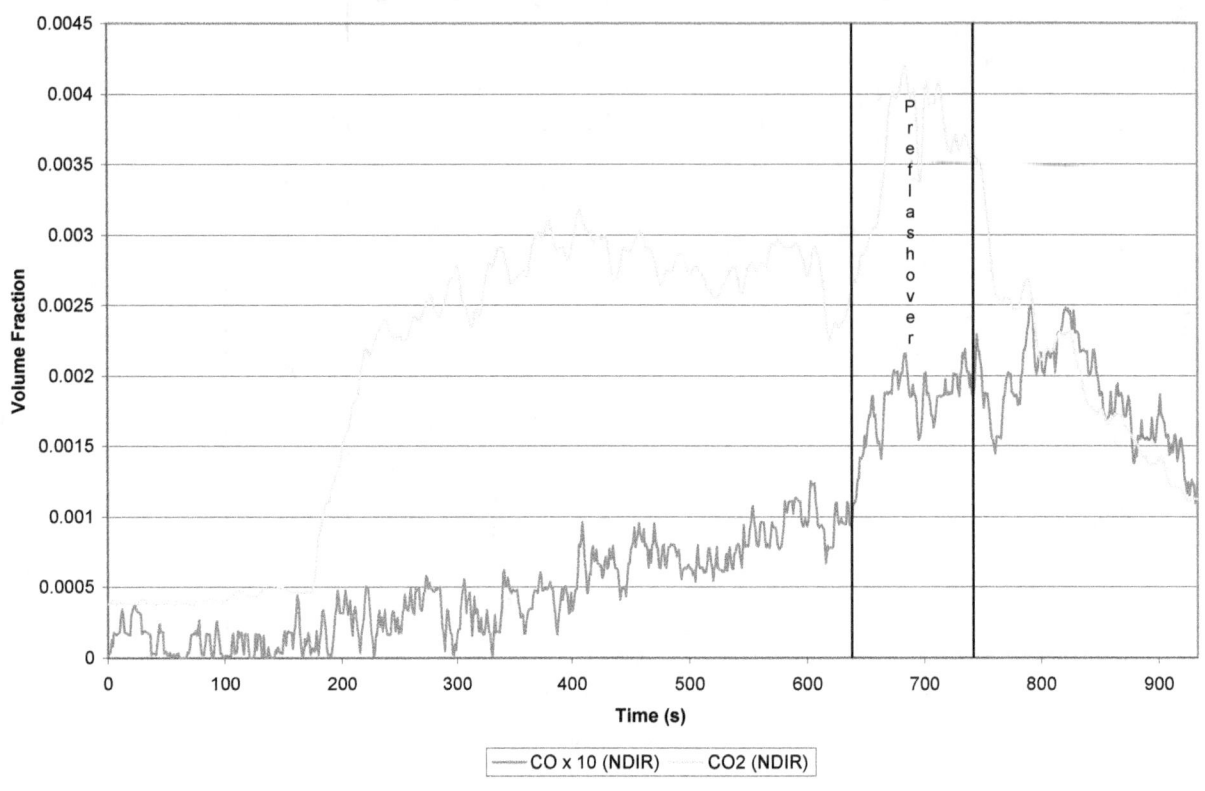

CO x 10 (NDIR)   CO2 (NDIR)

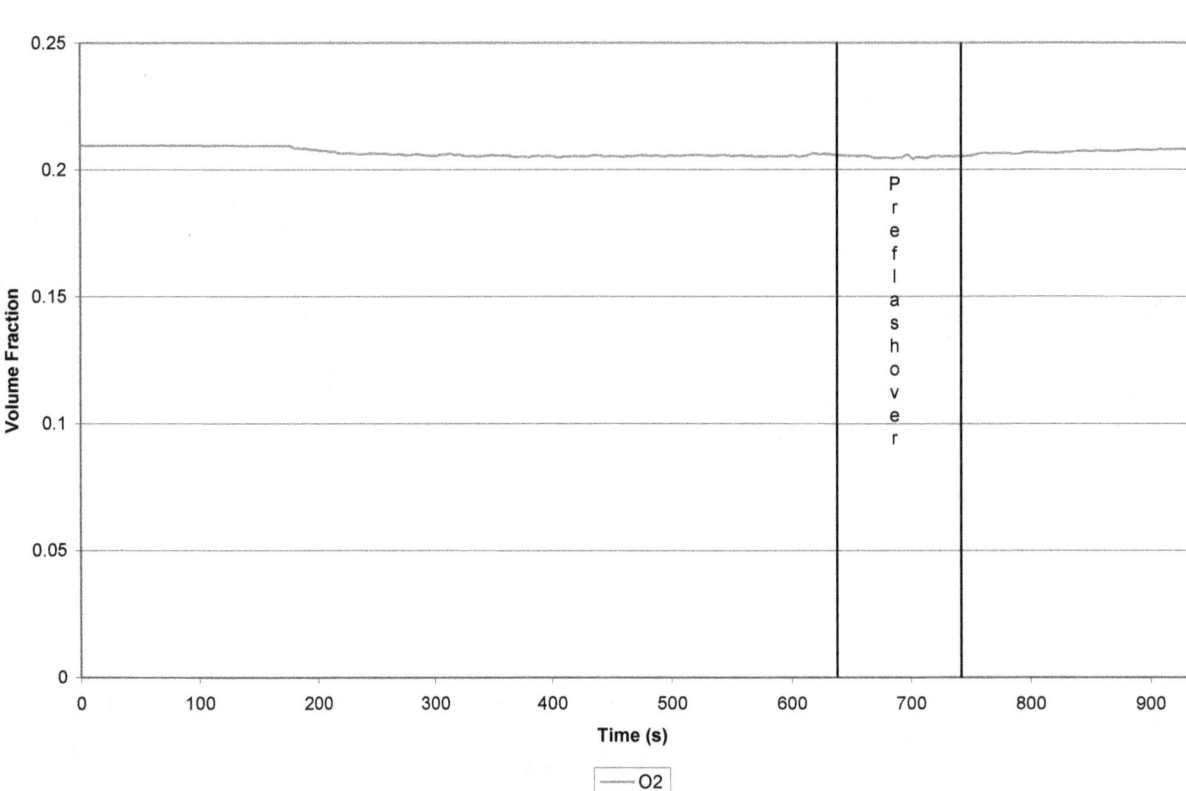

O2

Figures A14a, A14b. Data from Test BW6, Location 2

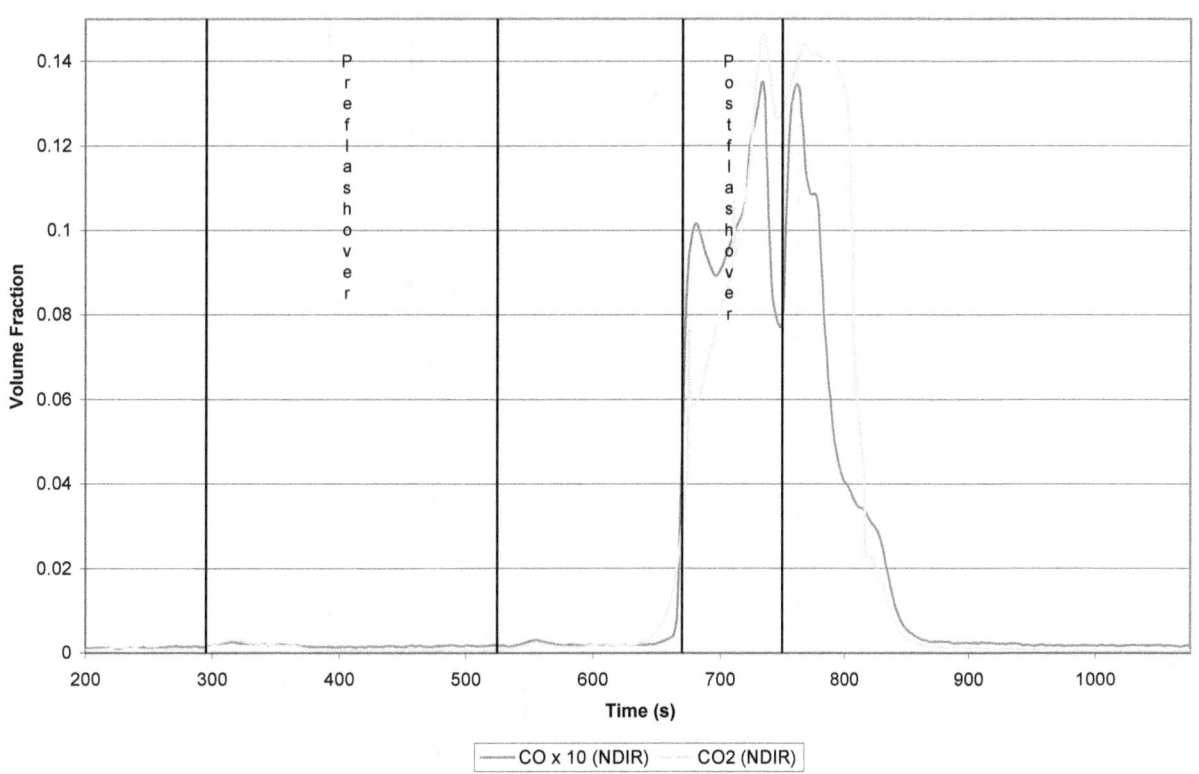

CO x 10 (NDIR)   CO2 (NDIR)

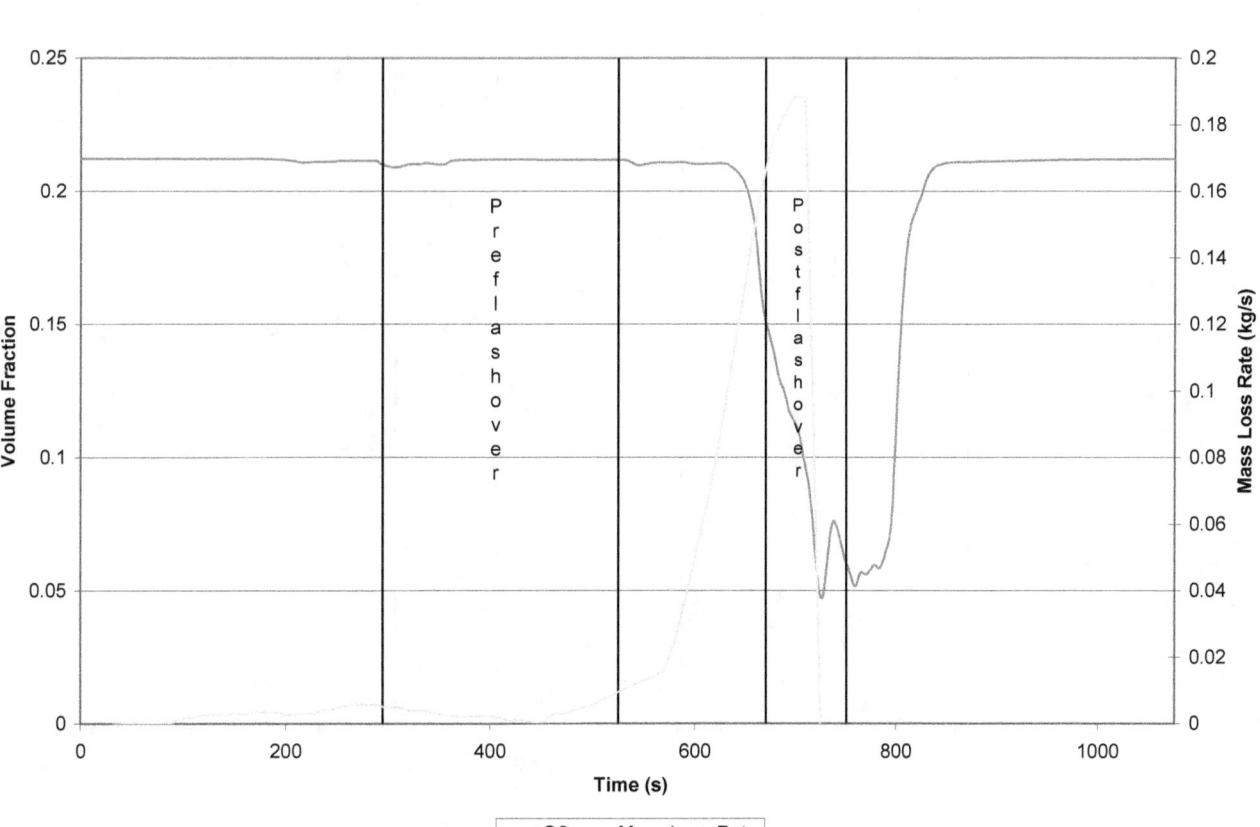

O2   Mass Loss Rate

Figures A14c, A14d. Data from Test BW6, Location 4

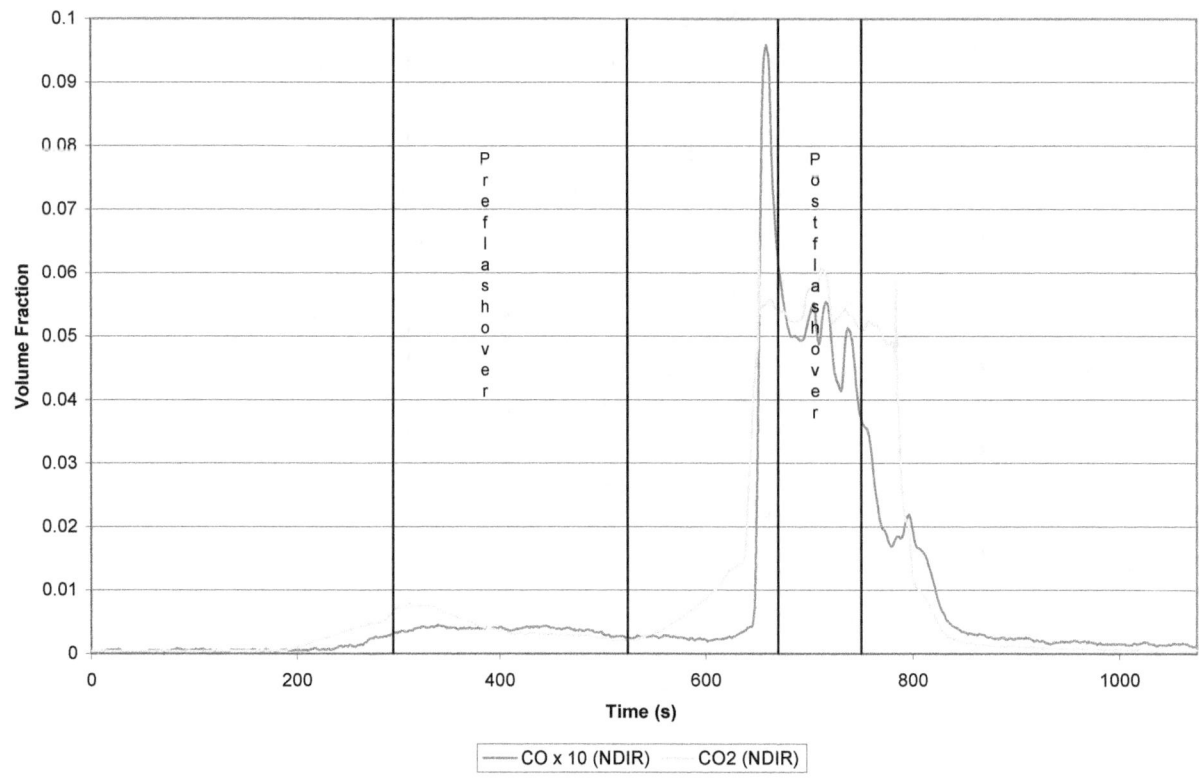

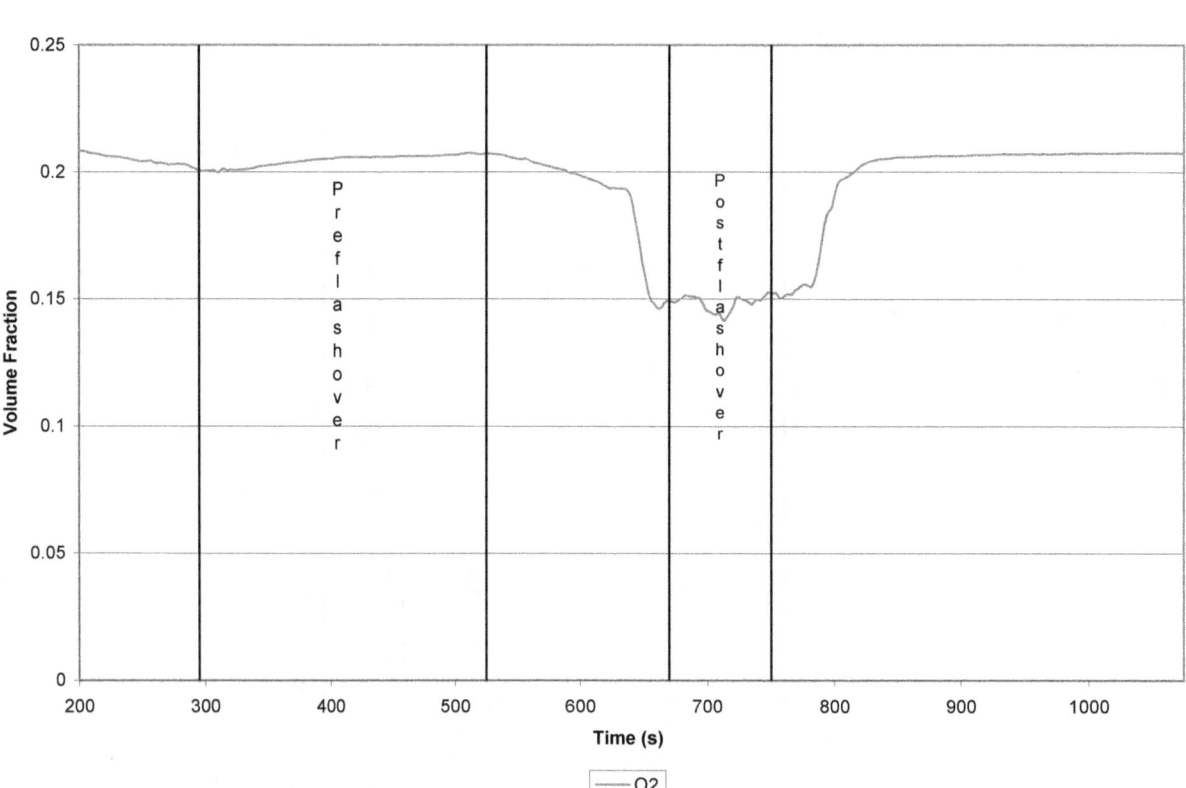

Figures A15a, A15b. Data from Test BW7, Location 2

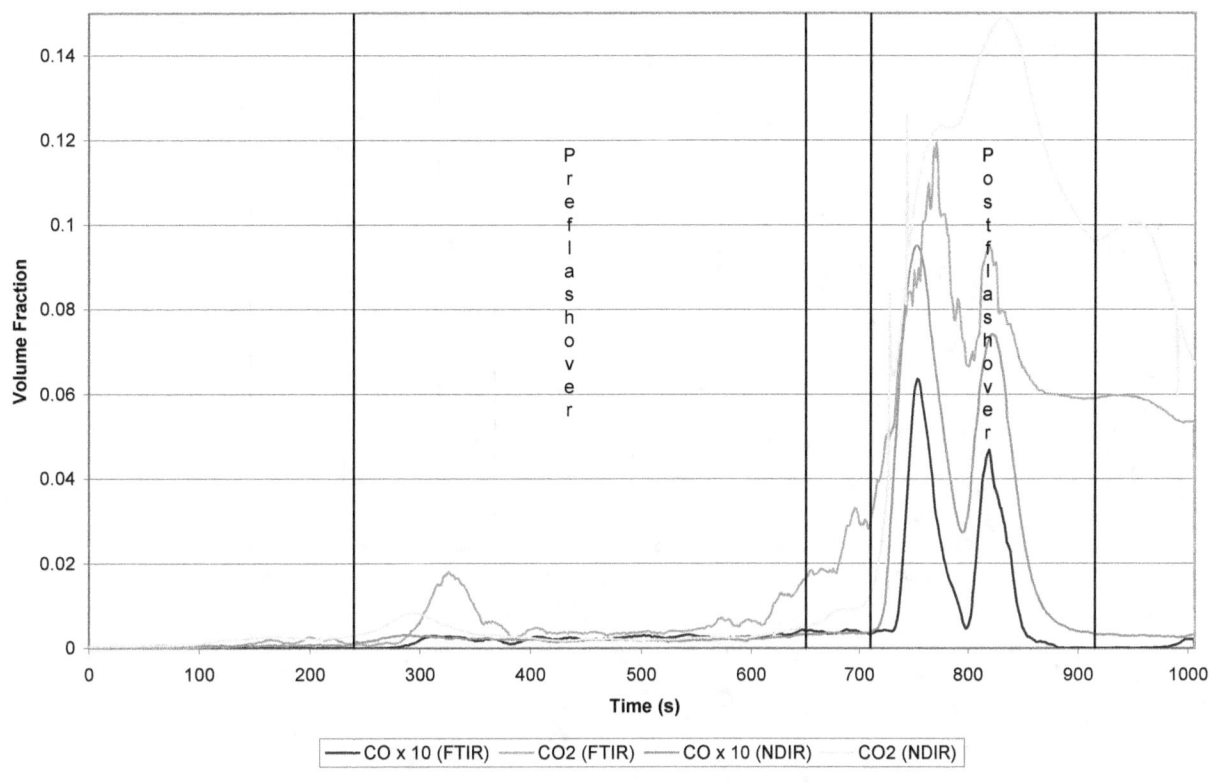

CO x 10 (FTIR) — CO2 (FTIR) — CO x 10 (NDIR) — CO2 (NDIR)

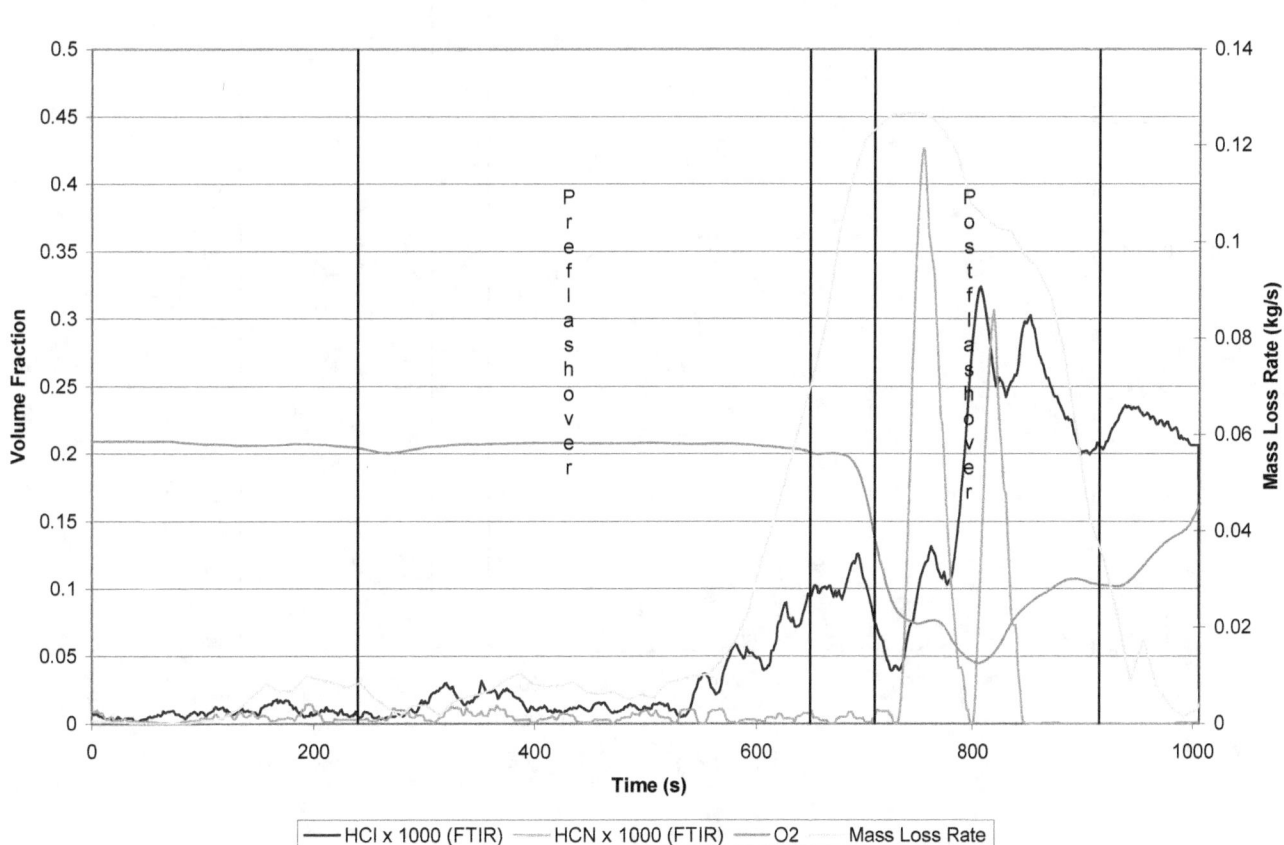

HCl x 1000 (FTIR) — HCN x 1000 (FTIR) — O2 — Mass Loss Rate

Figures A15c, A15d. Data from Test BW7, Location 4

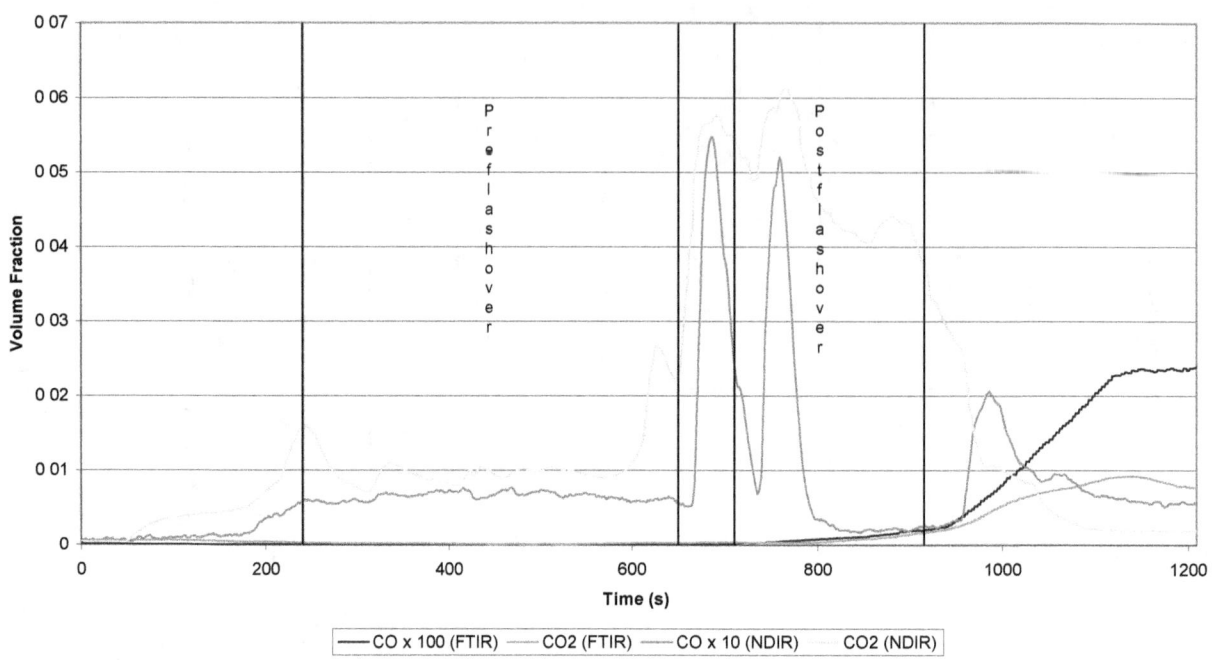

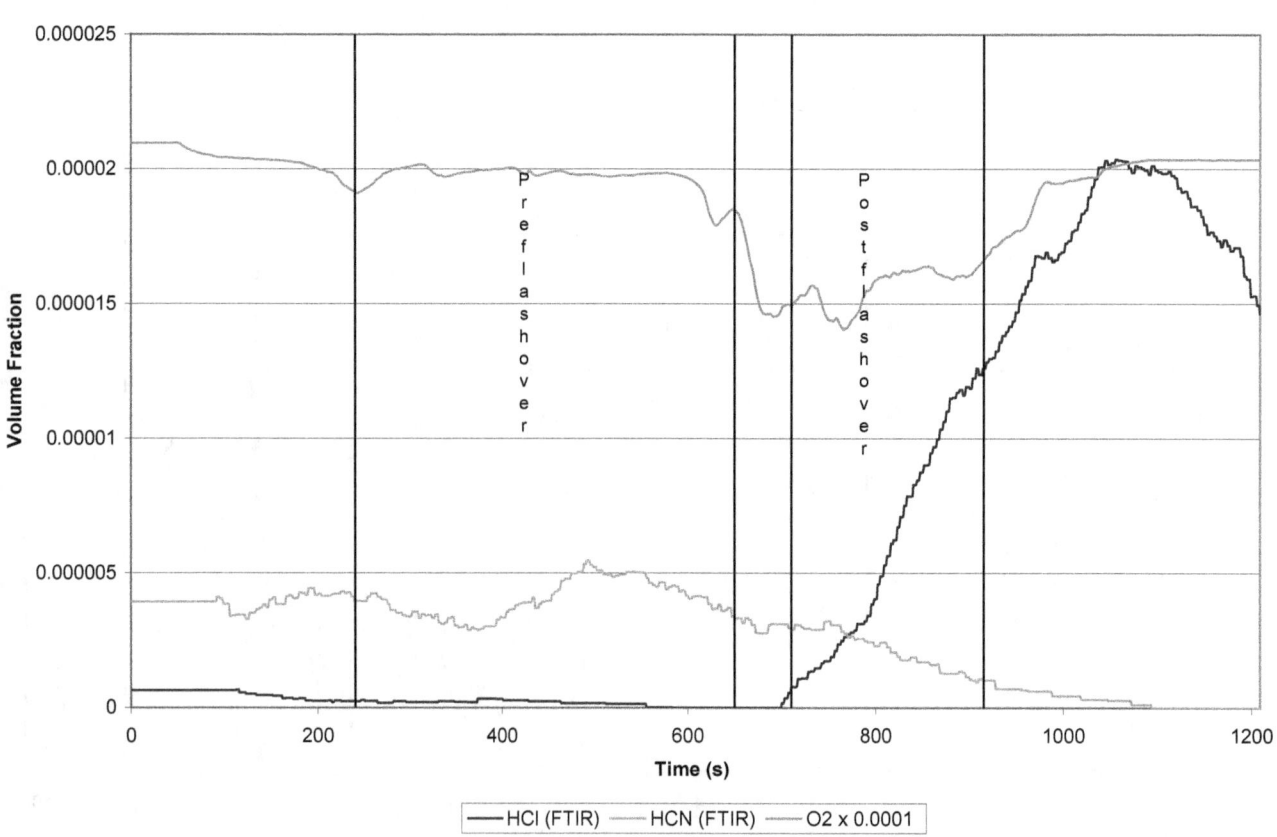

Figures A16a, A16b. Data from Test BP1, Location 2

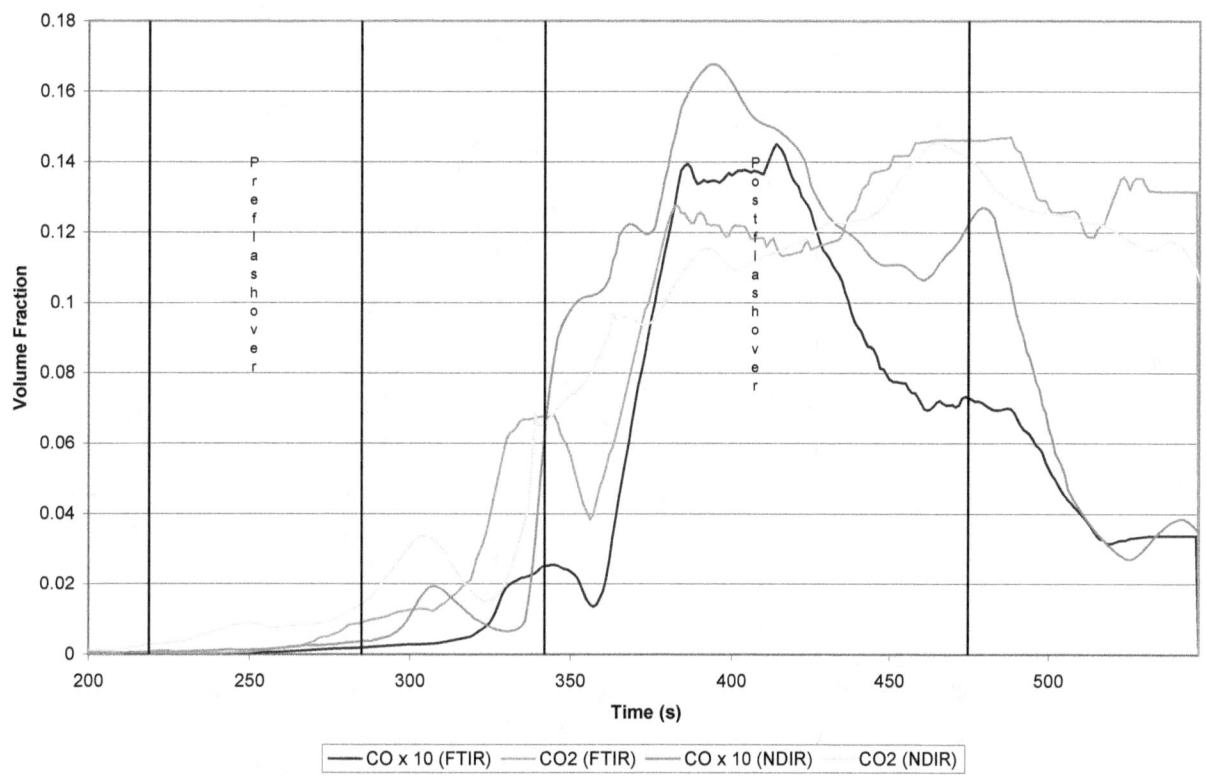

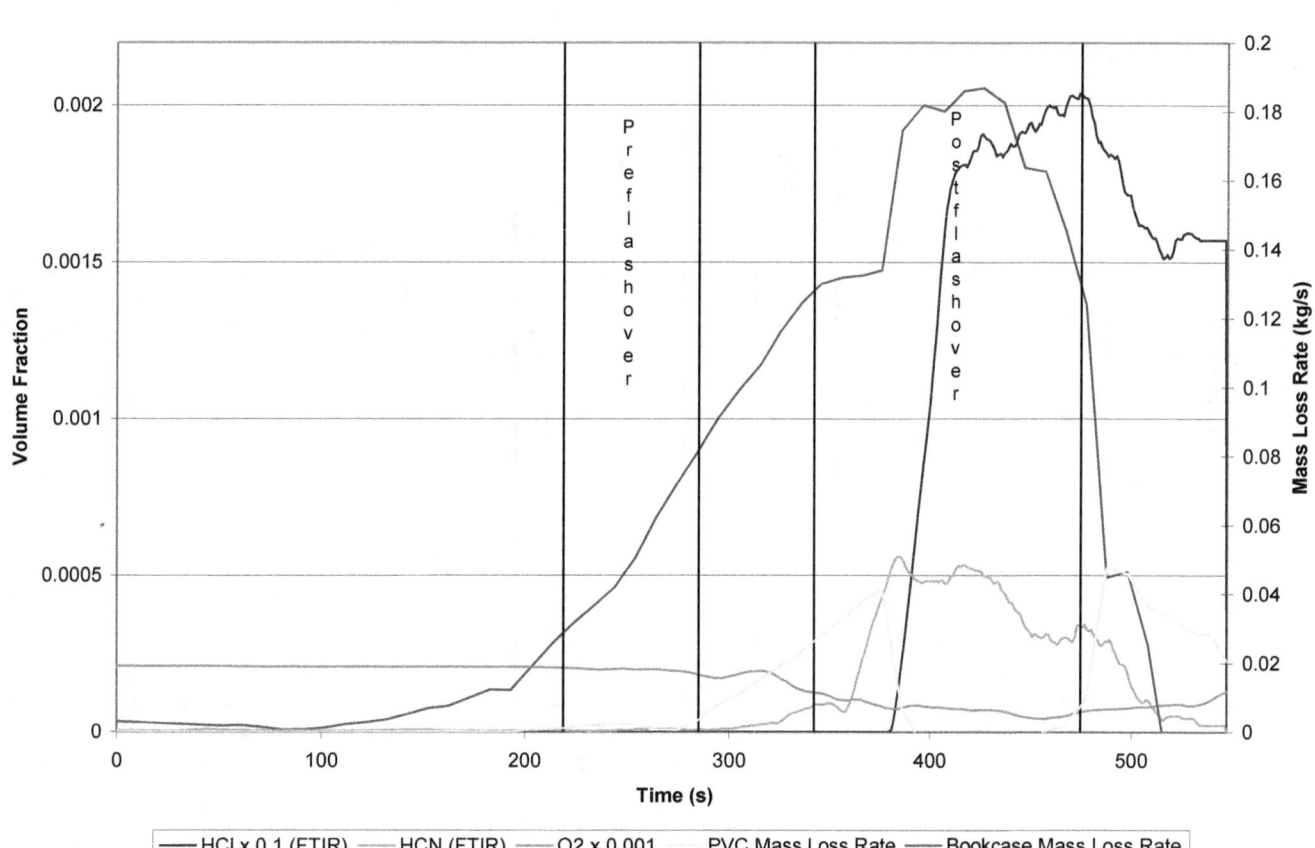

Figures A16c, A16d. Data from Test BP1, Location 4

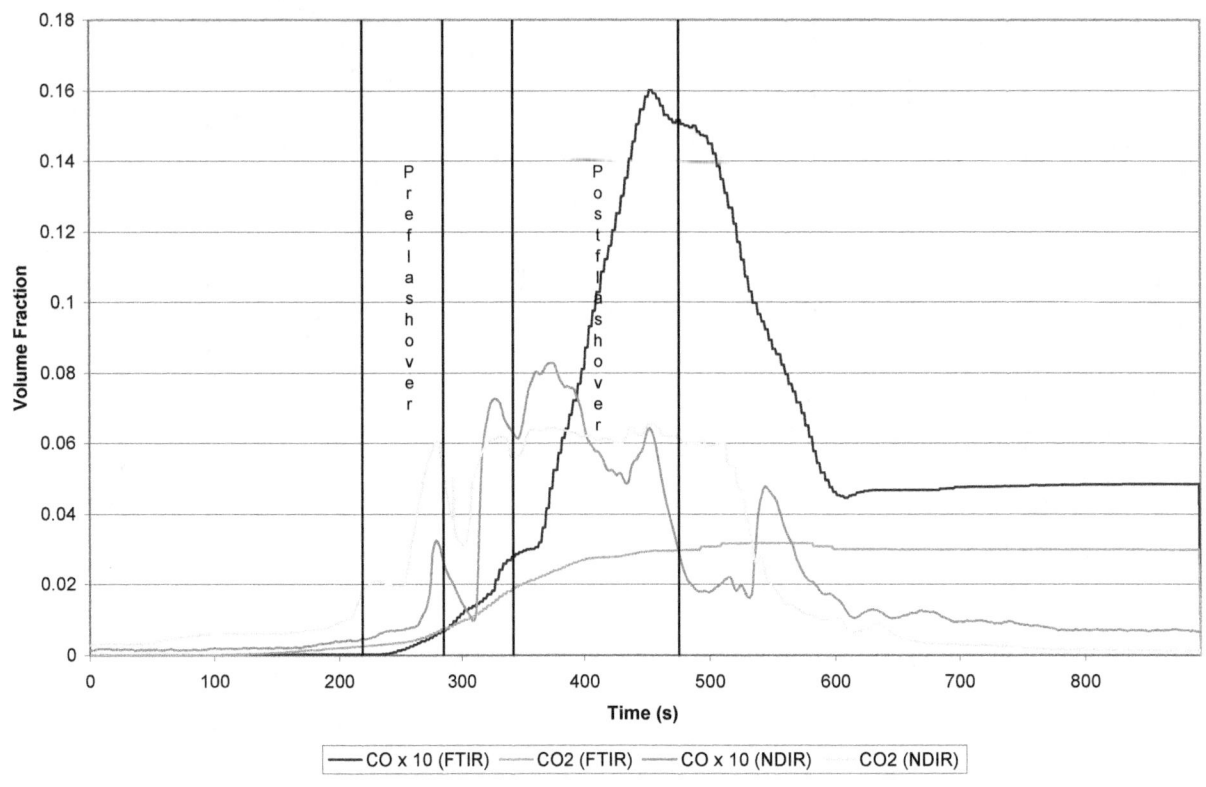

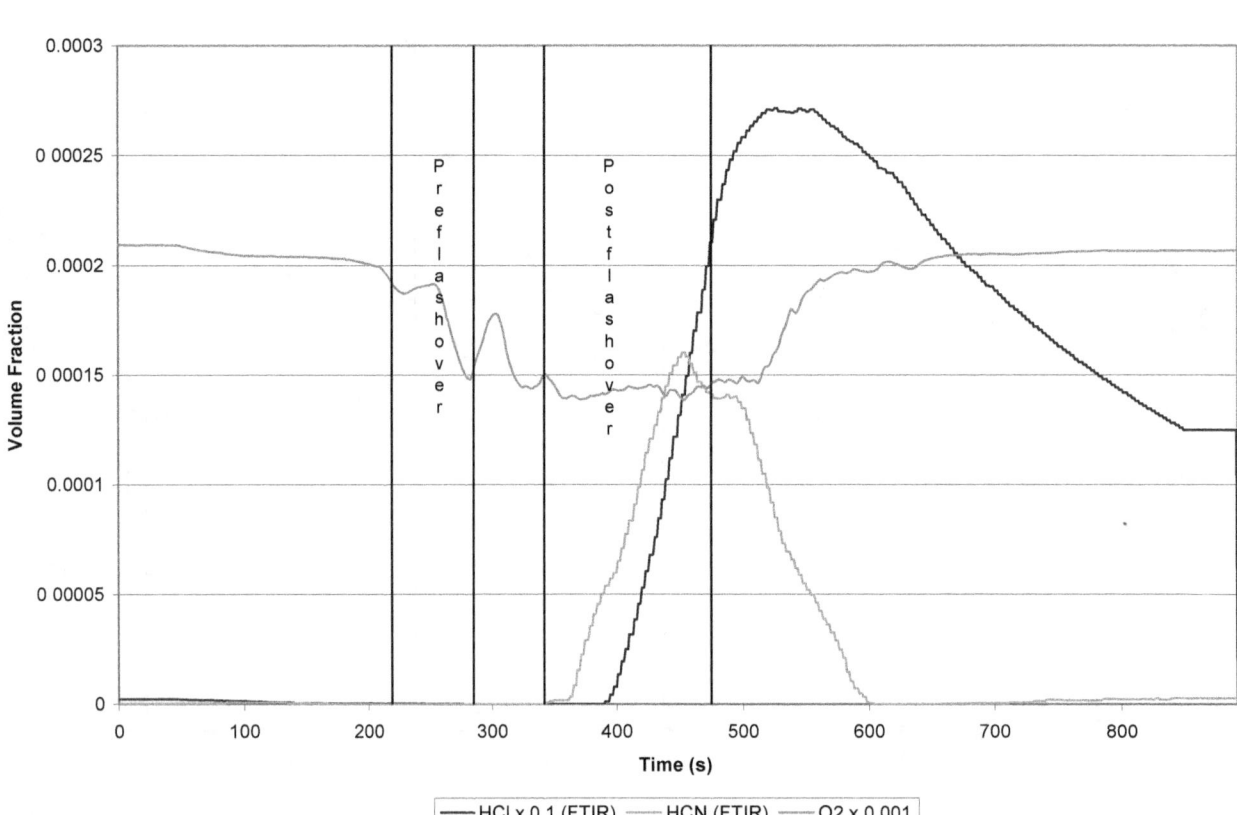

Figures A17a, A17b. Data from Test BP2, Location 2

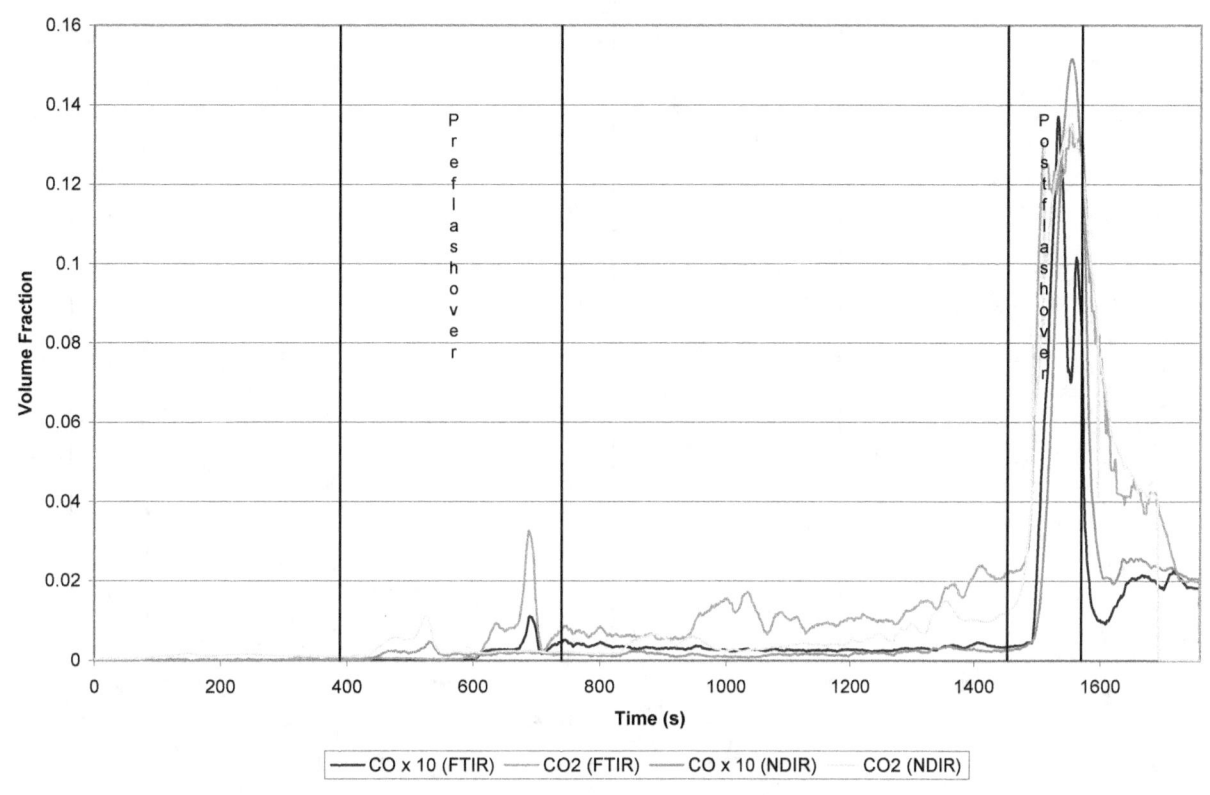

CO x 10 (FTIR) — CO2 (FTIR) — CO x 10 (NDIR) — CO2 (NDIR)

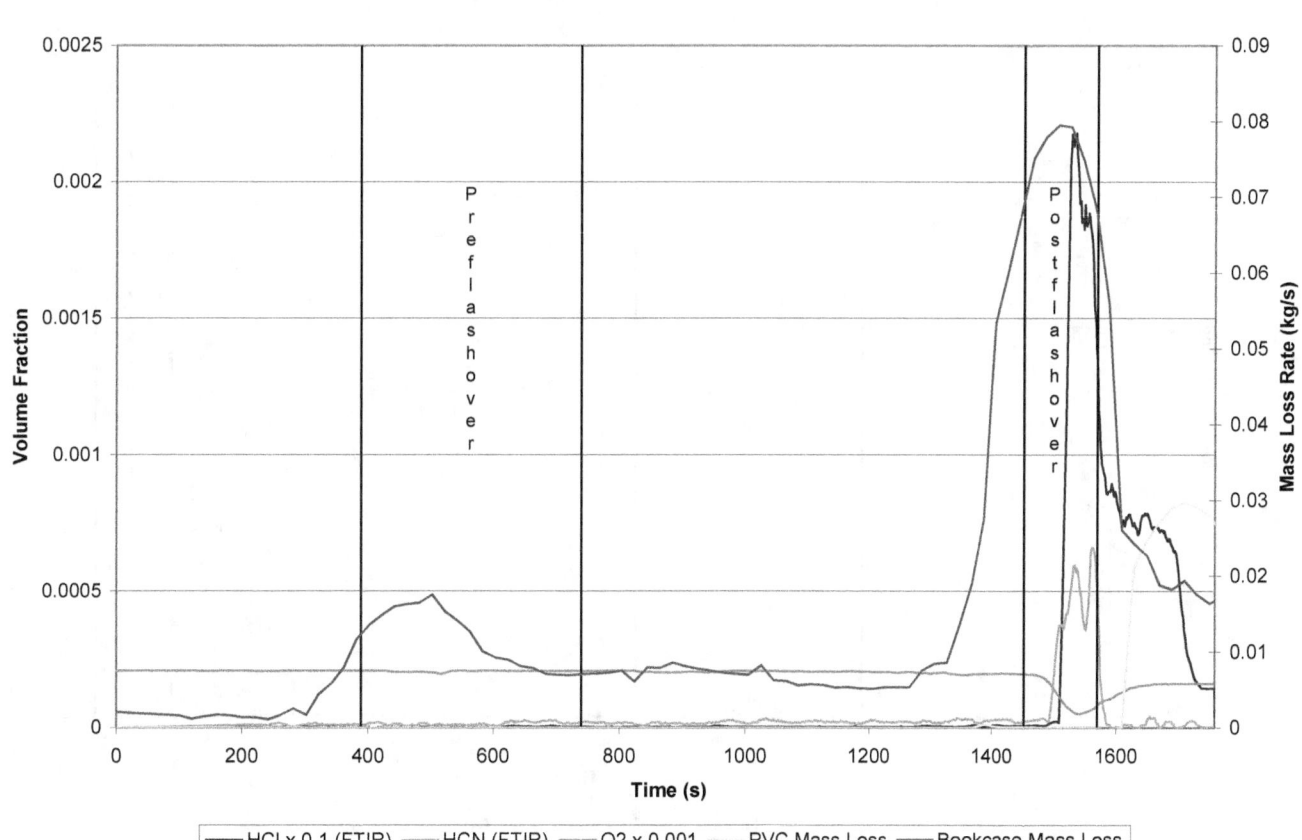

HCl x 0.1 (FTIR) — HCN (FTIR) — O2 x 0.001 — PVC Mass Loss — Bookcase Mass Loss

Figures A17c, A17d. Data from Test BP2, Location 4

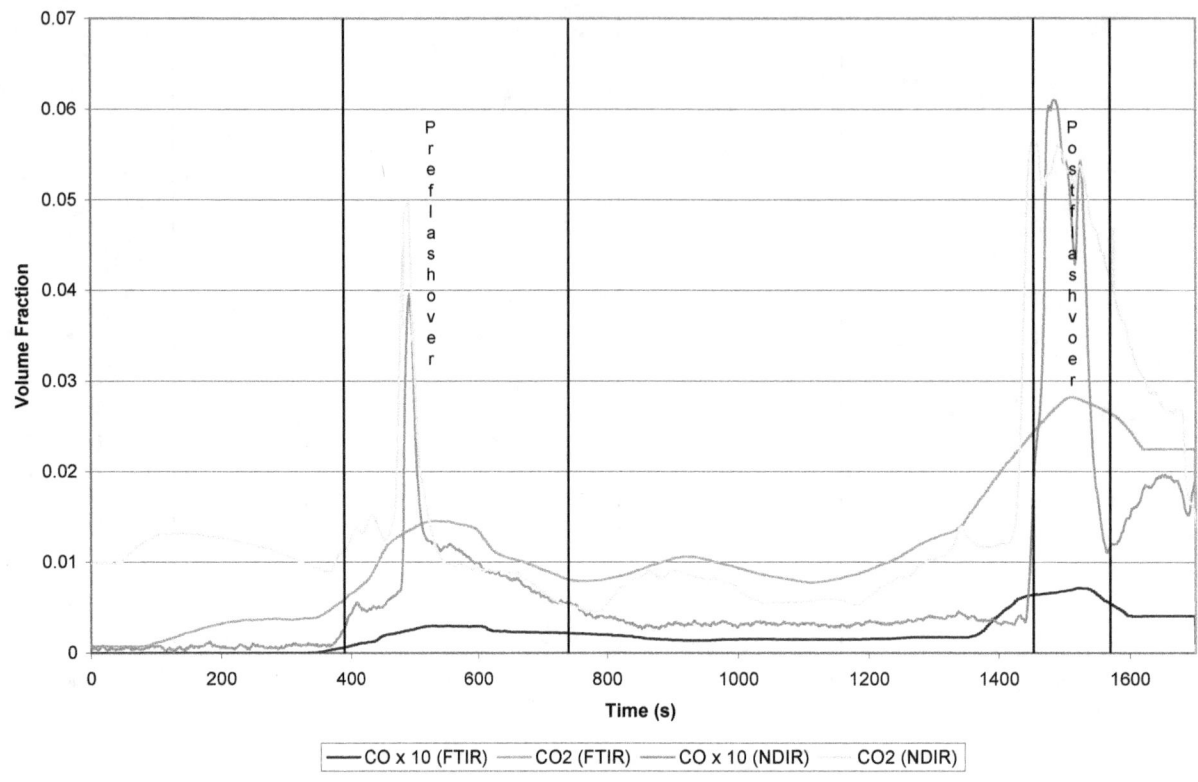

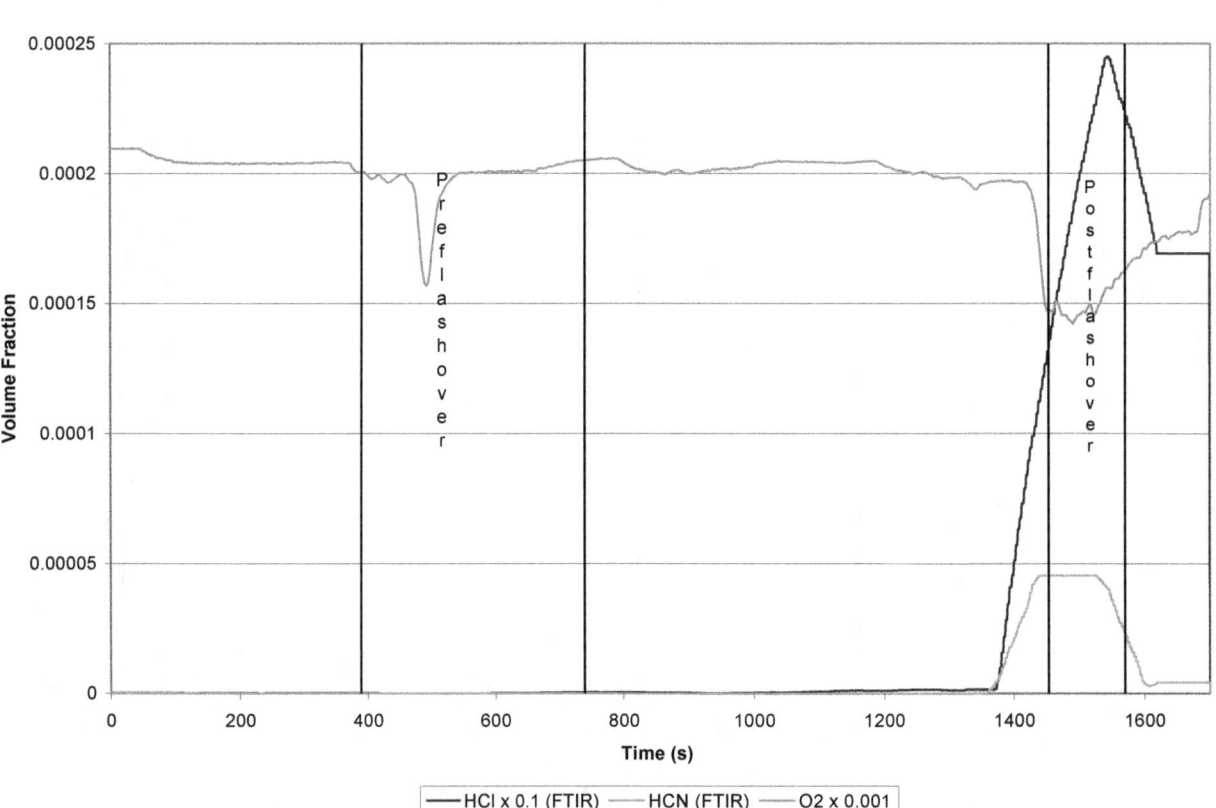

Figures A18a, A18b. Data from Test BP3, Location 2

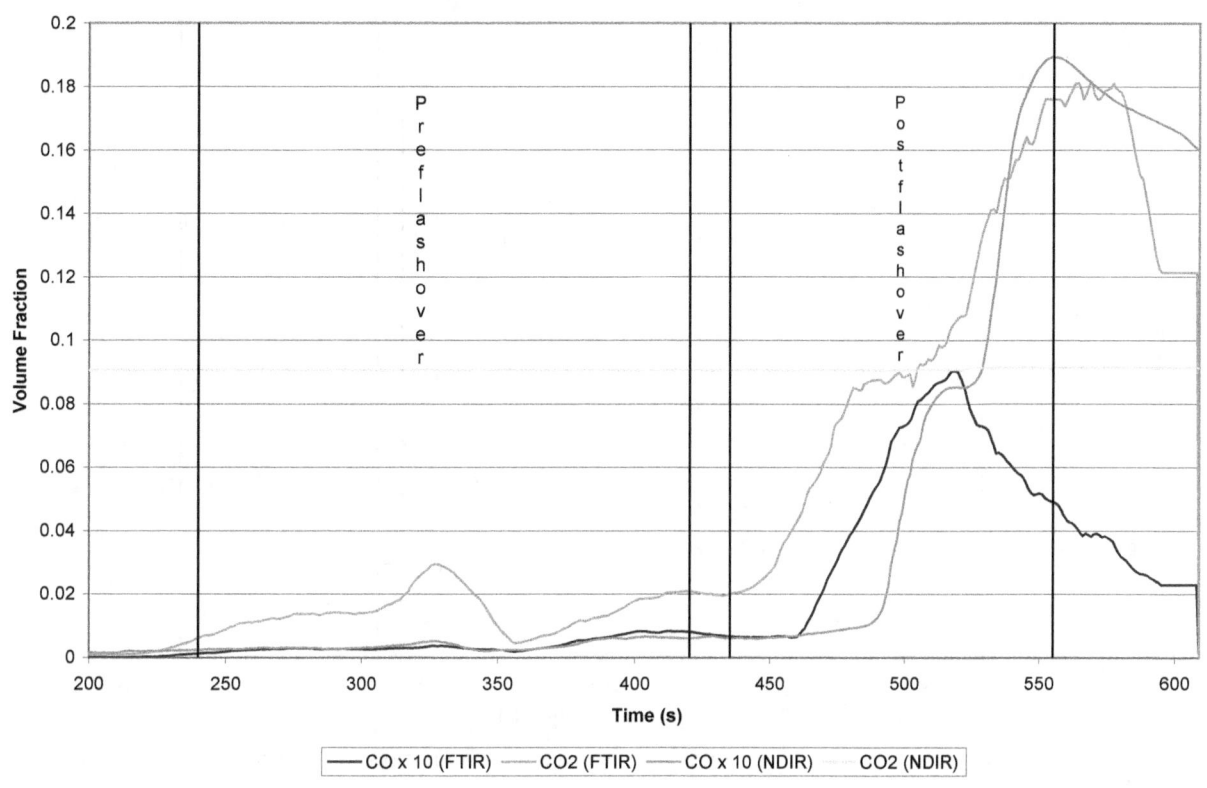

CO x 10 (FTIR) —— CO2 (FTIR) —— CO x 10 (NDIR) —— CO2 (NDIR)

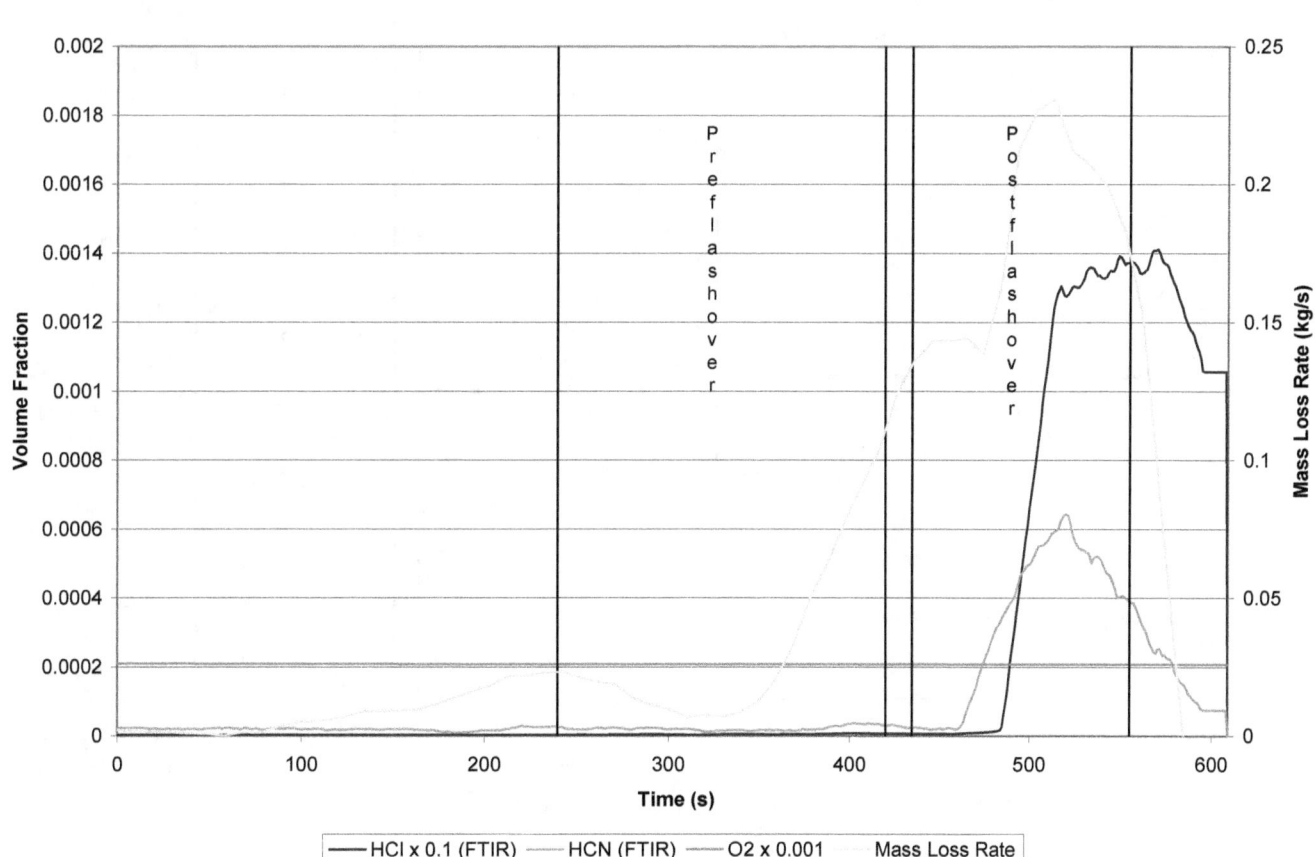

HCl x 0.1 (FTIR) —— HCN (FTIR) —— O2 x 0.001 —— Mass Loss Rate

Figures A18c, A18d. Data from Test BP3, Location 4

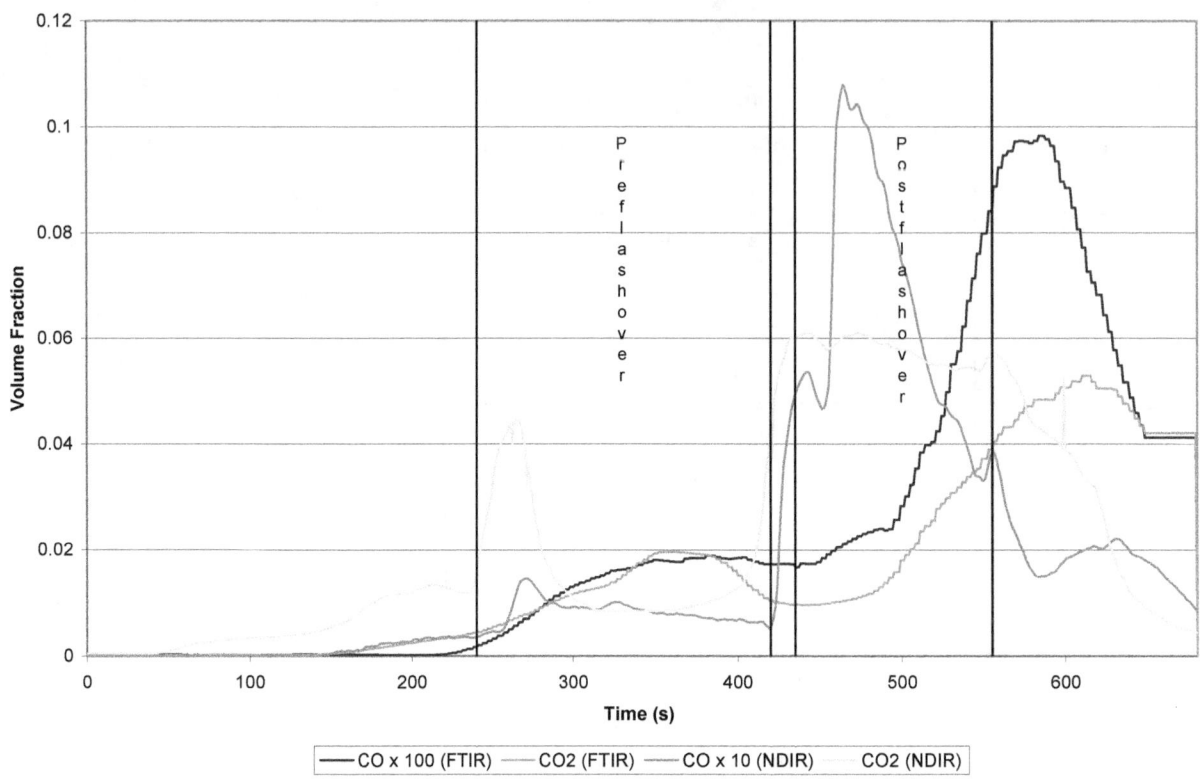

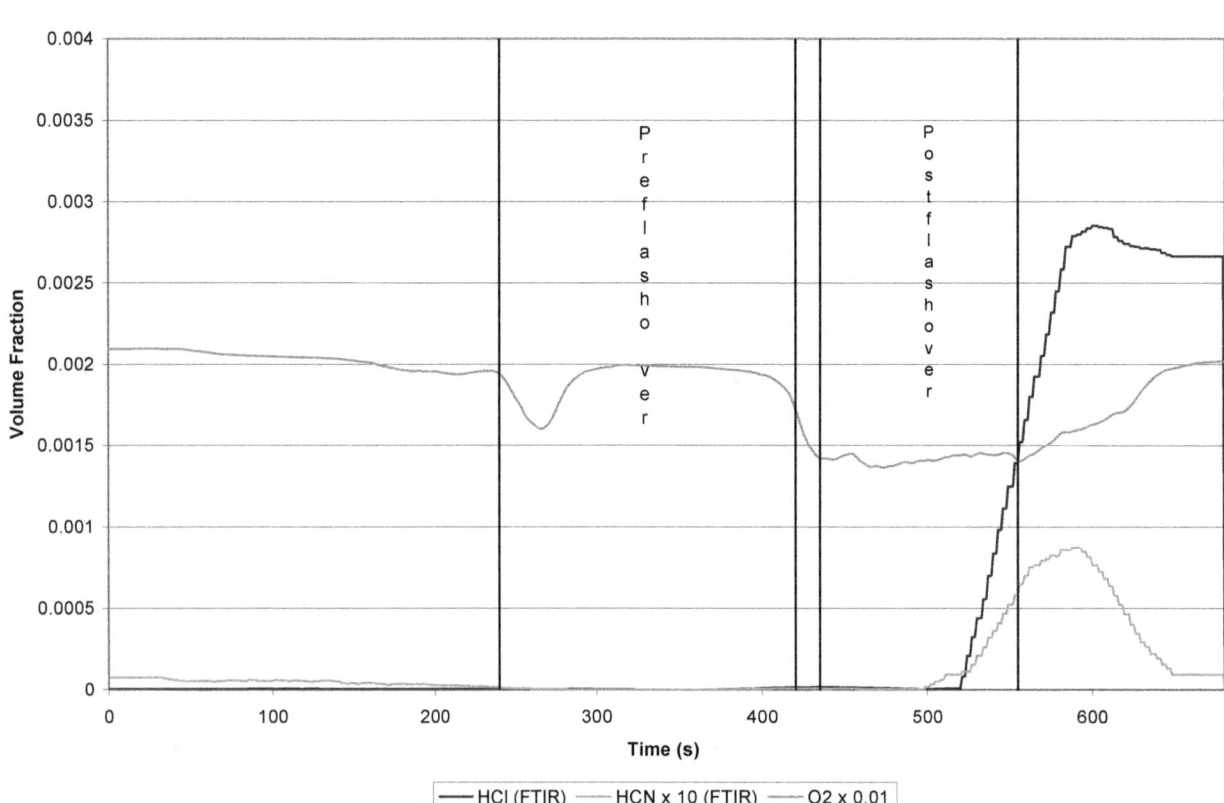

Figures A19a, A19b. Data from Test PQ1, Location 2

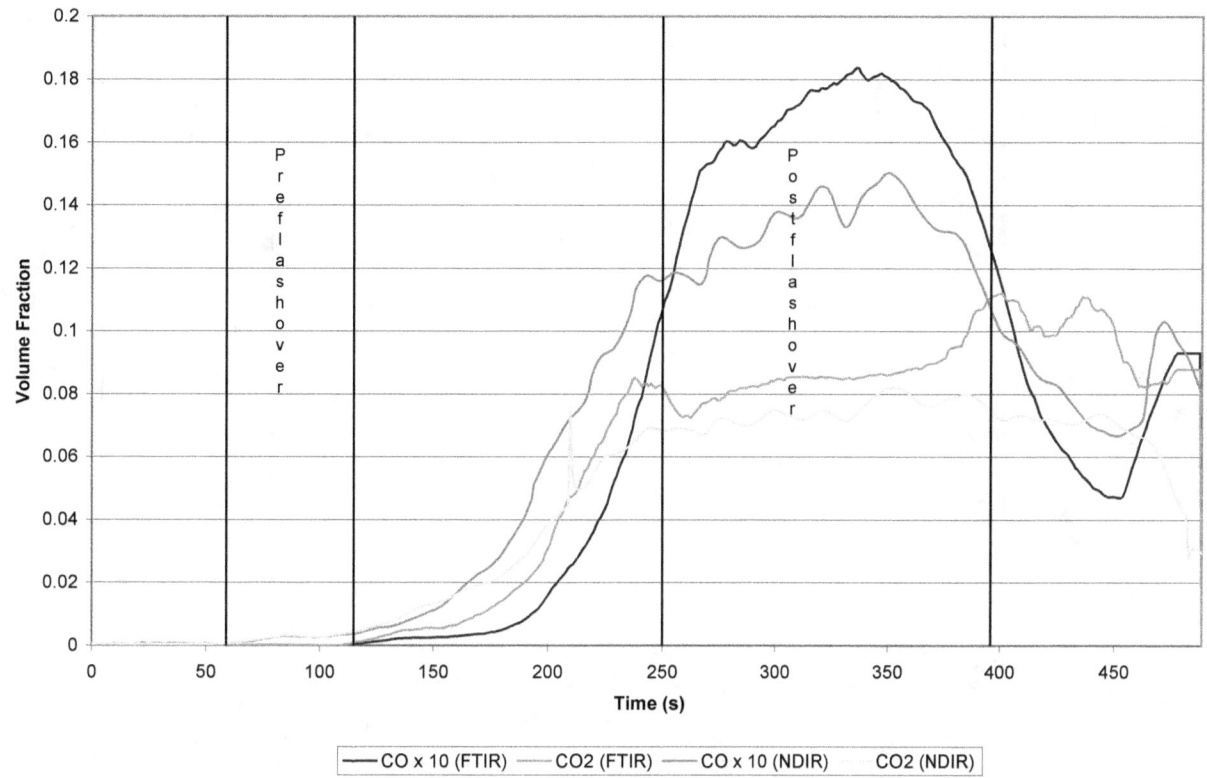

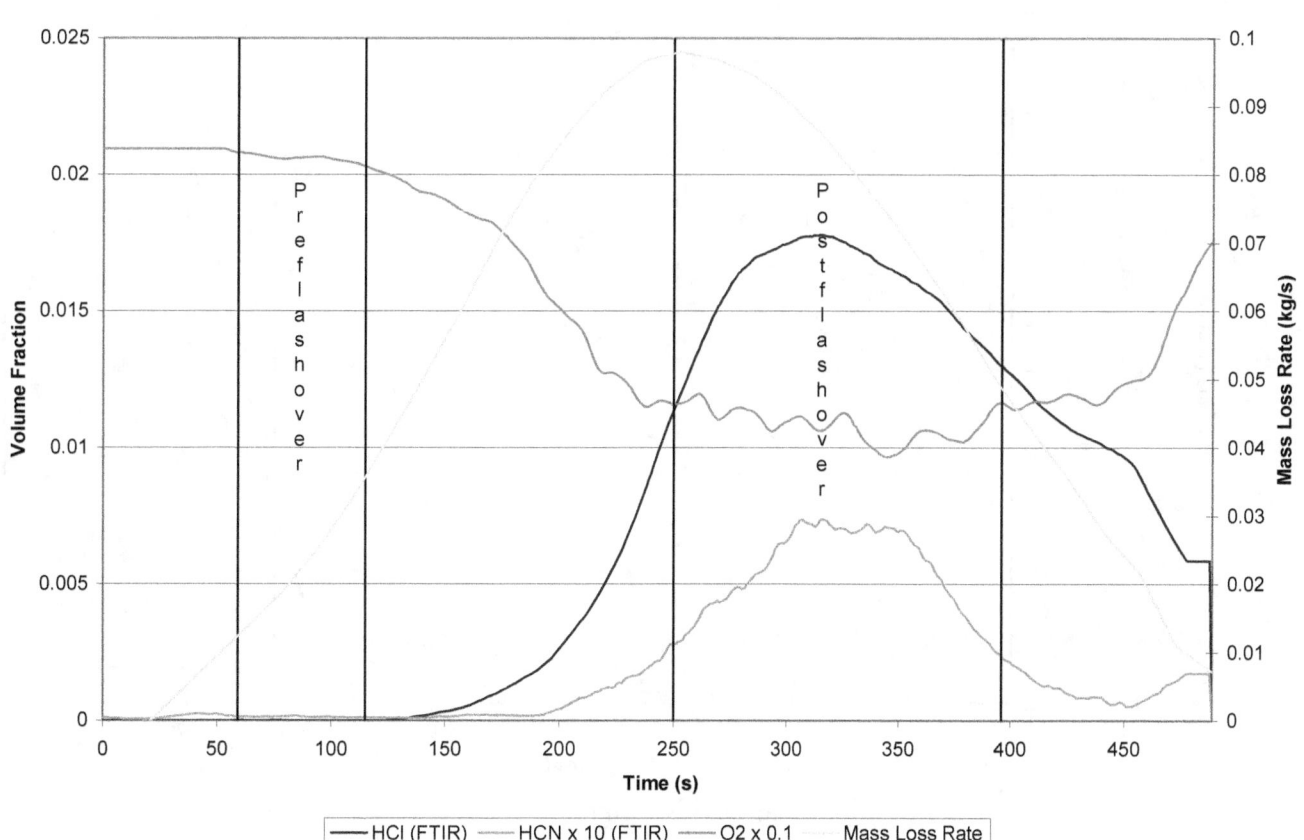

Figures A20a, A20b. Data from Test PQ2, Location 2

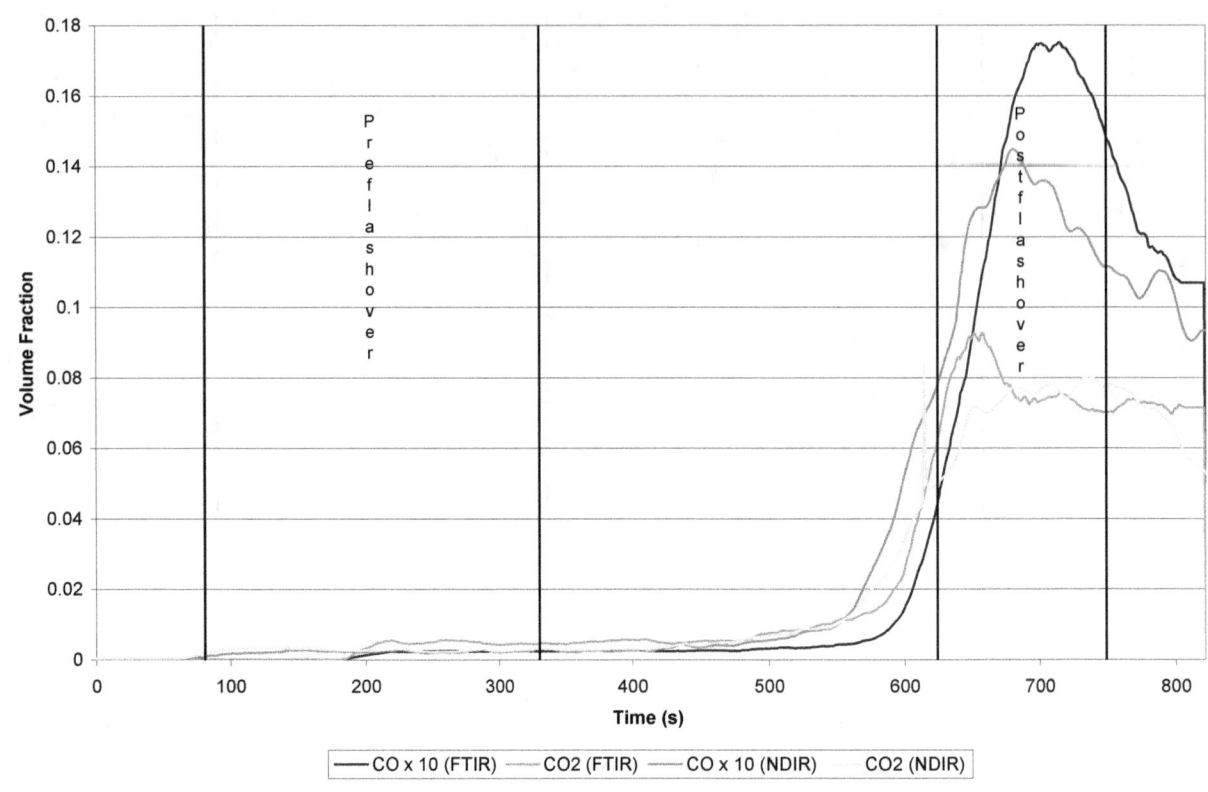

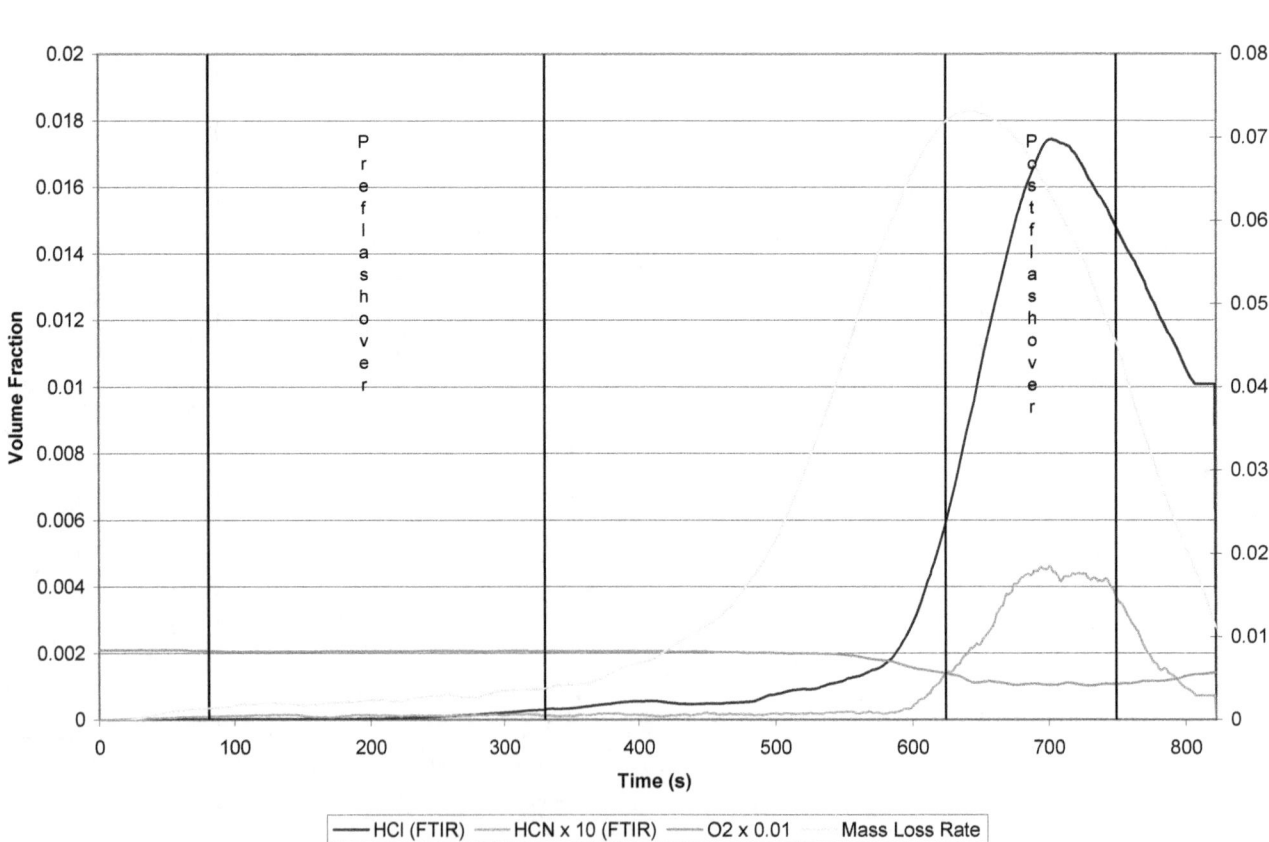

Figures A21a, A21b. Data from Test PW1, Location 2

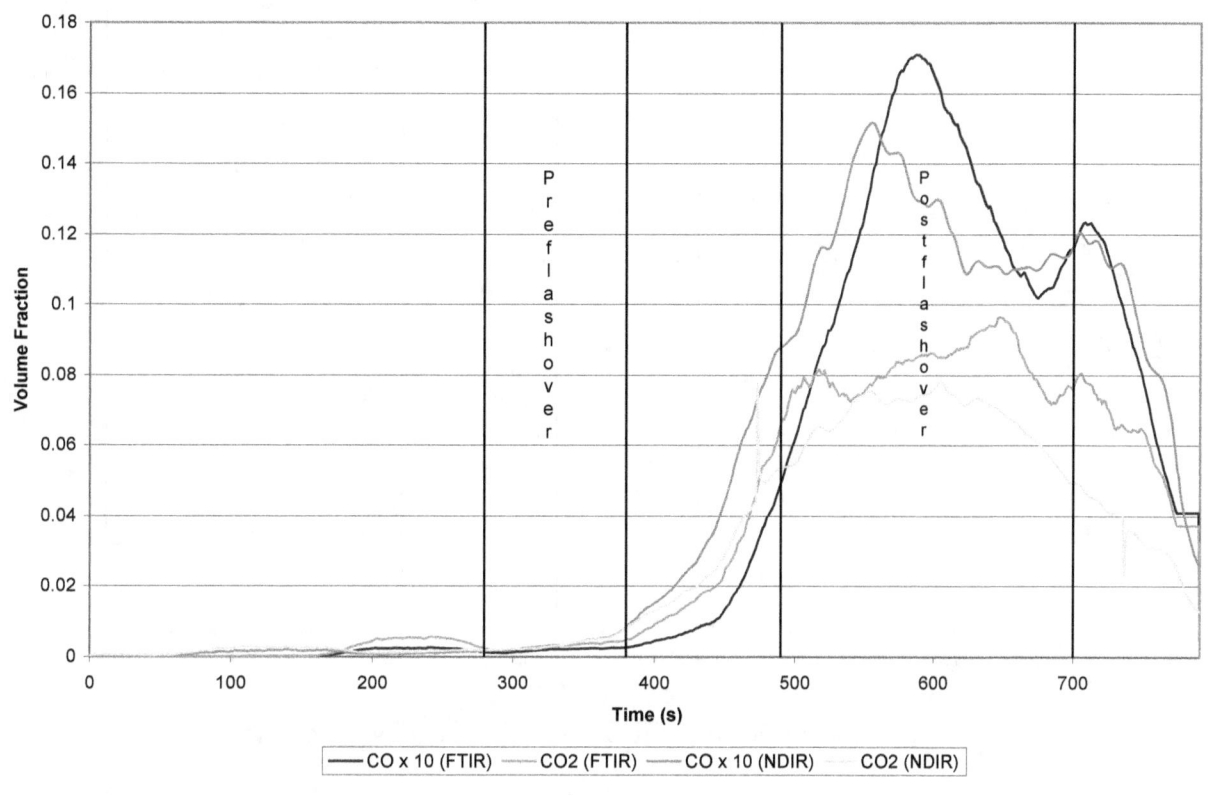

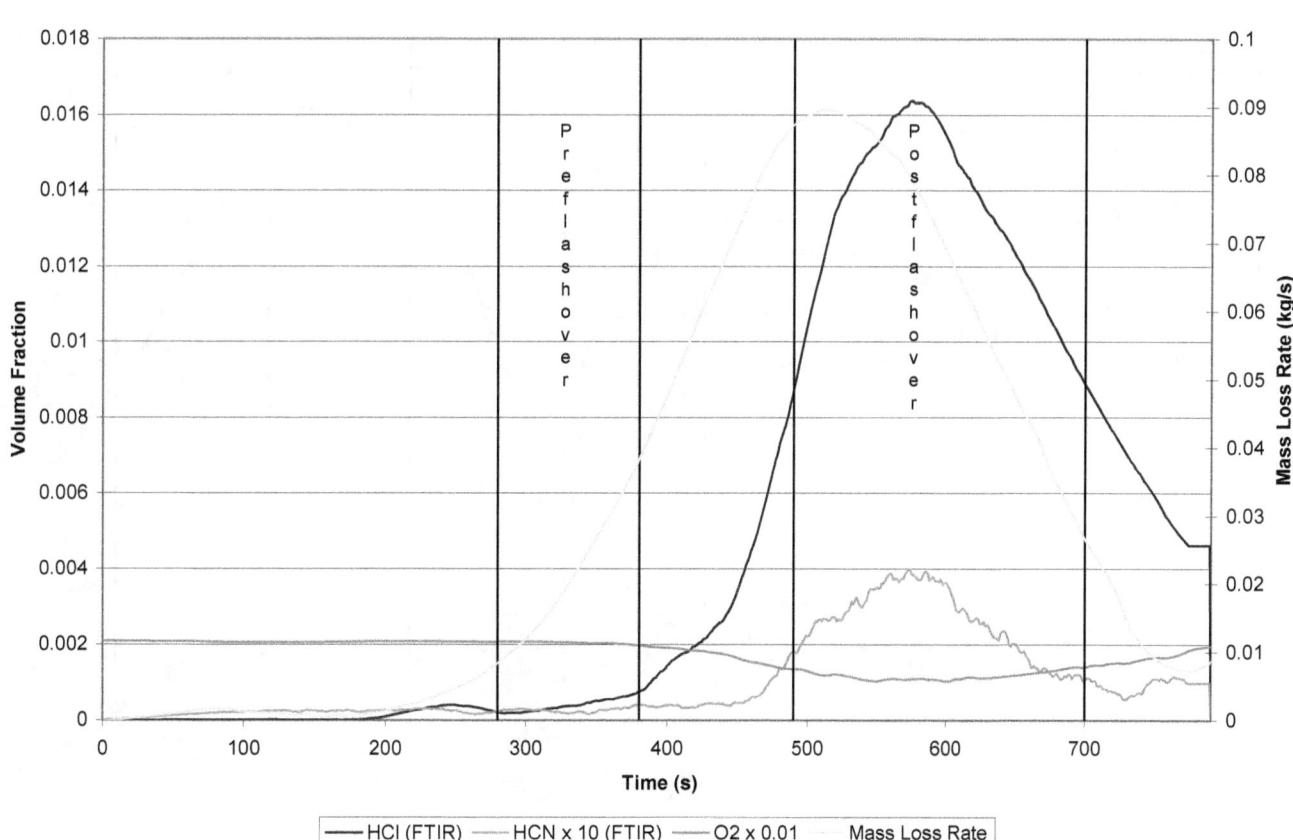

Figures A21c, A21d. Data from Test PW1, Location 4

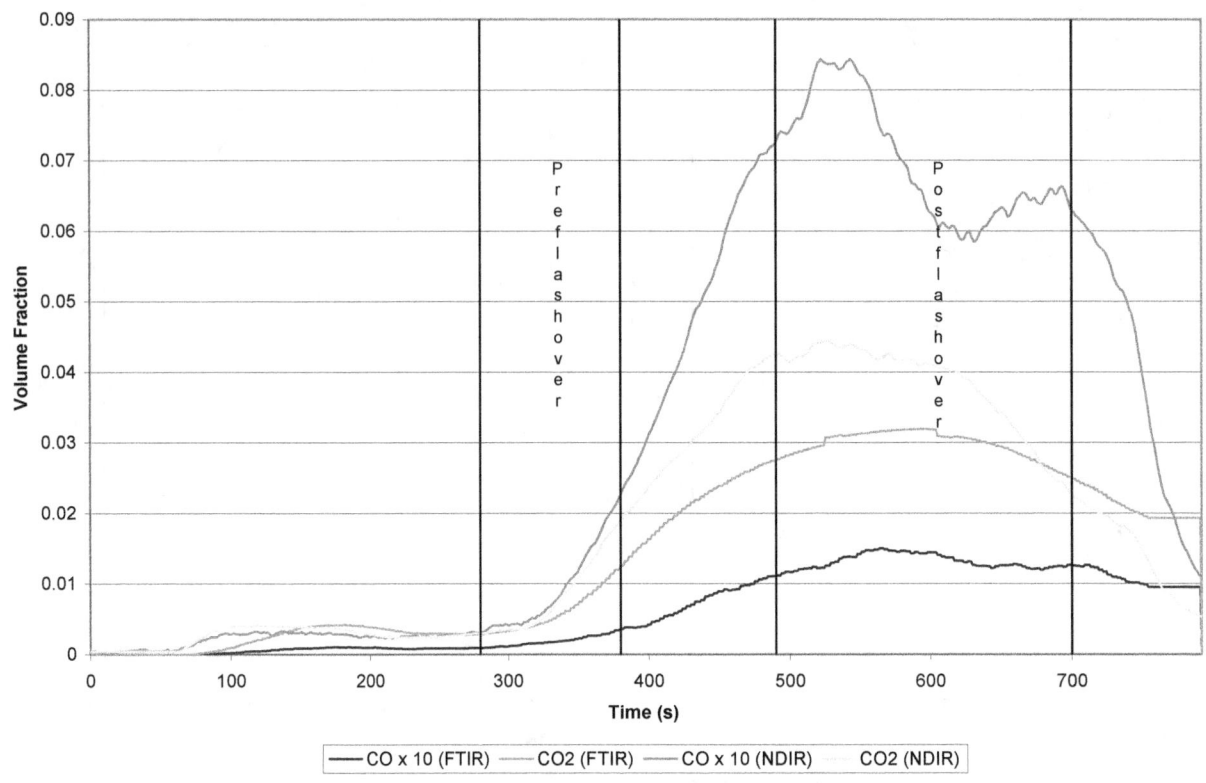

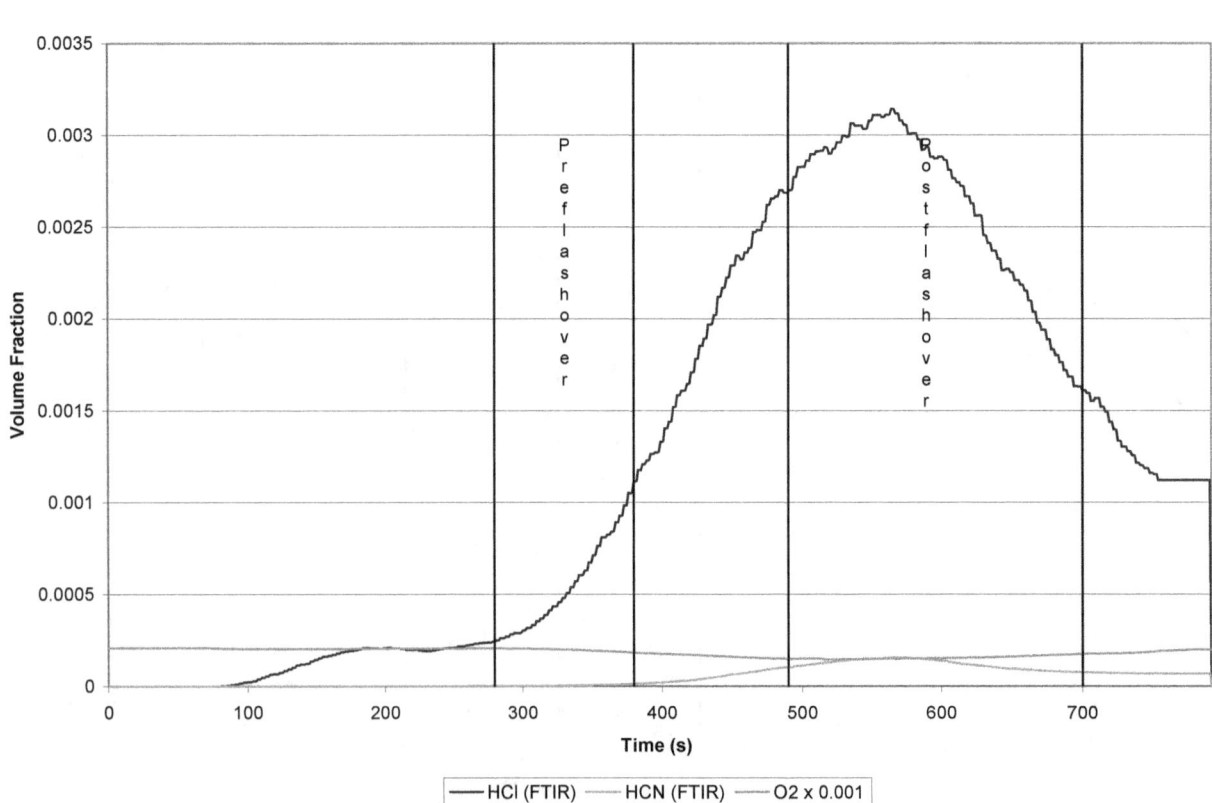

Figures A22a, A22b. Data from Test PW2, Location 2

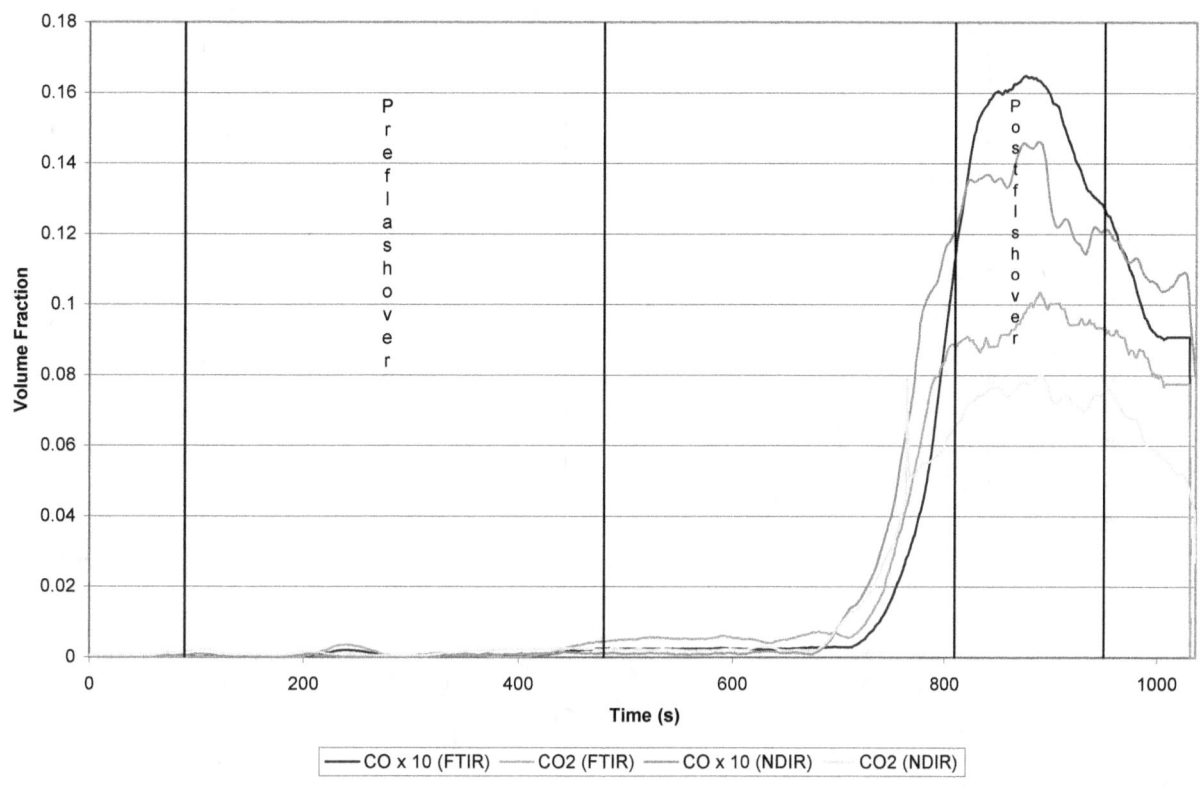

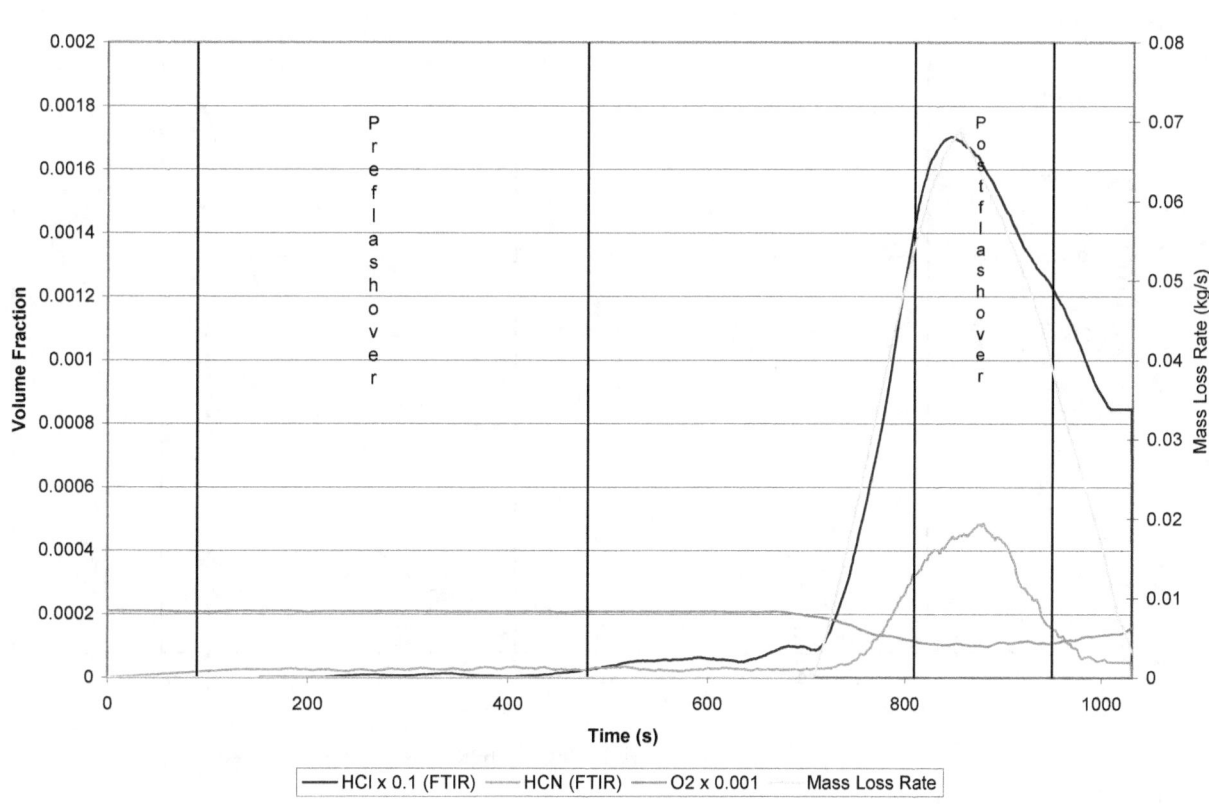

Figures A22c, A22d. Data from Test PW2, Location 4

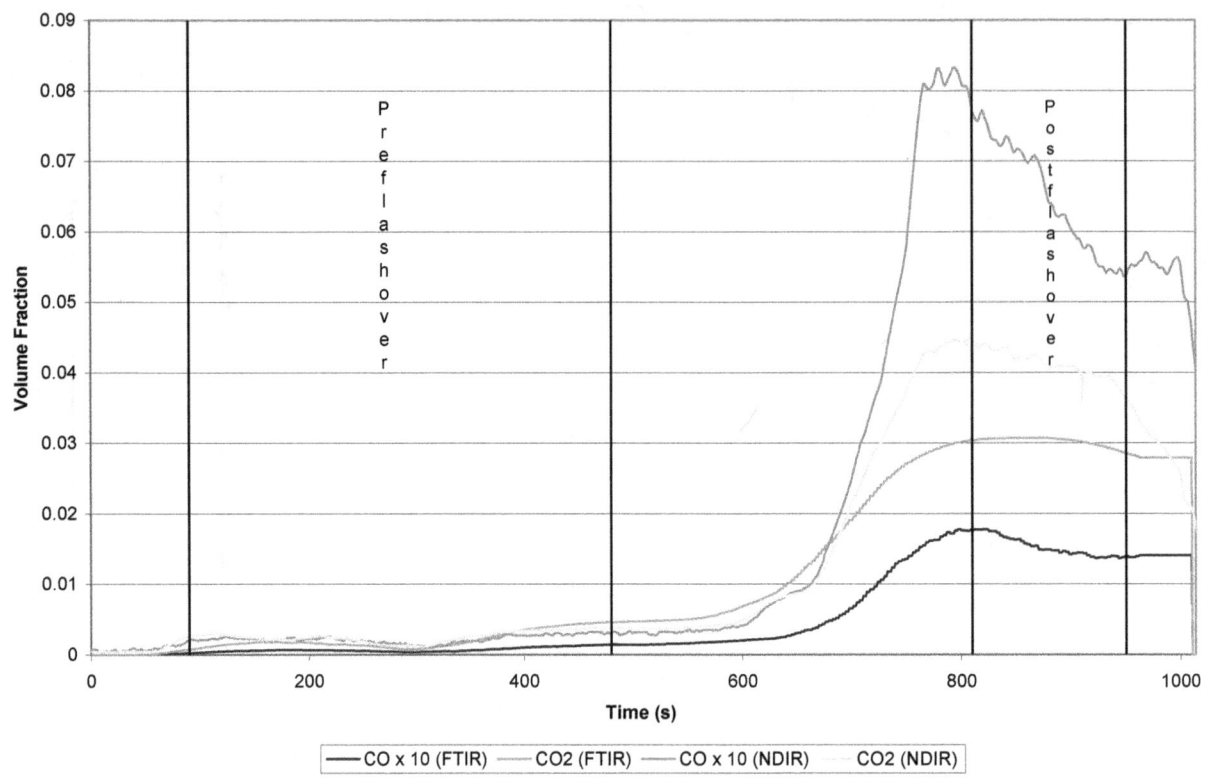

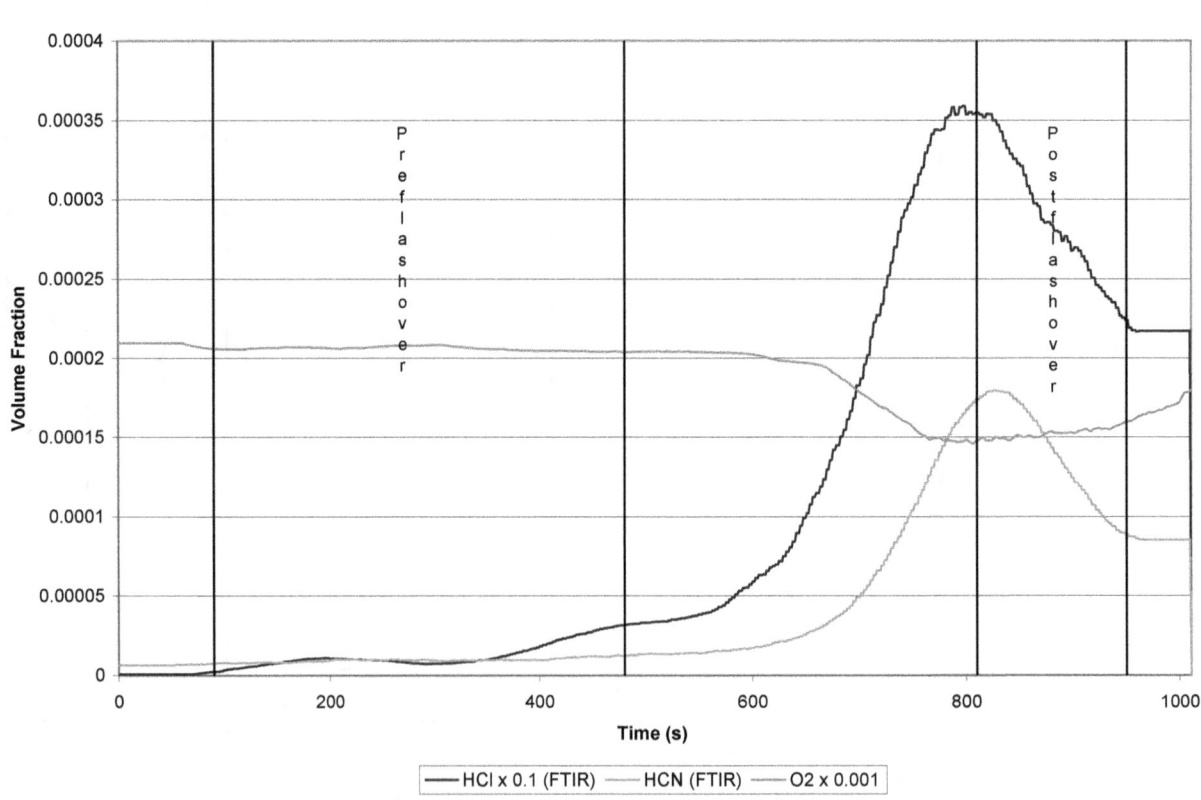

Figures A23a, A23b. Data from Test SC1, Location 1

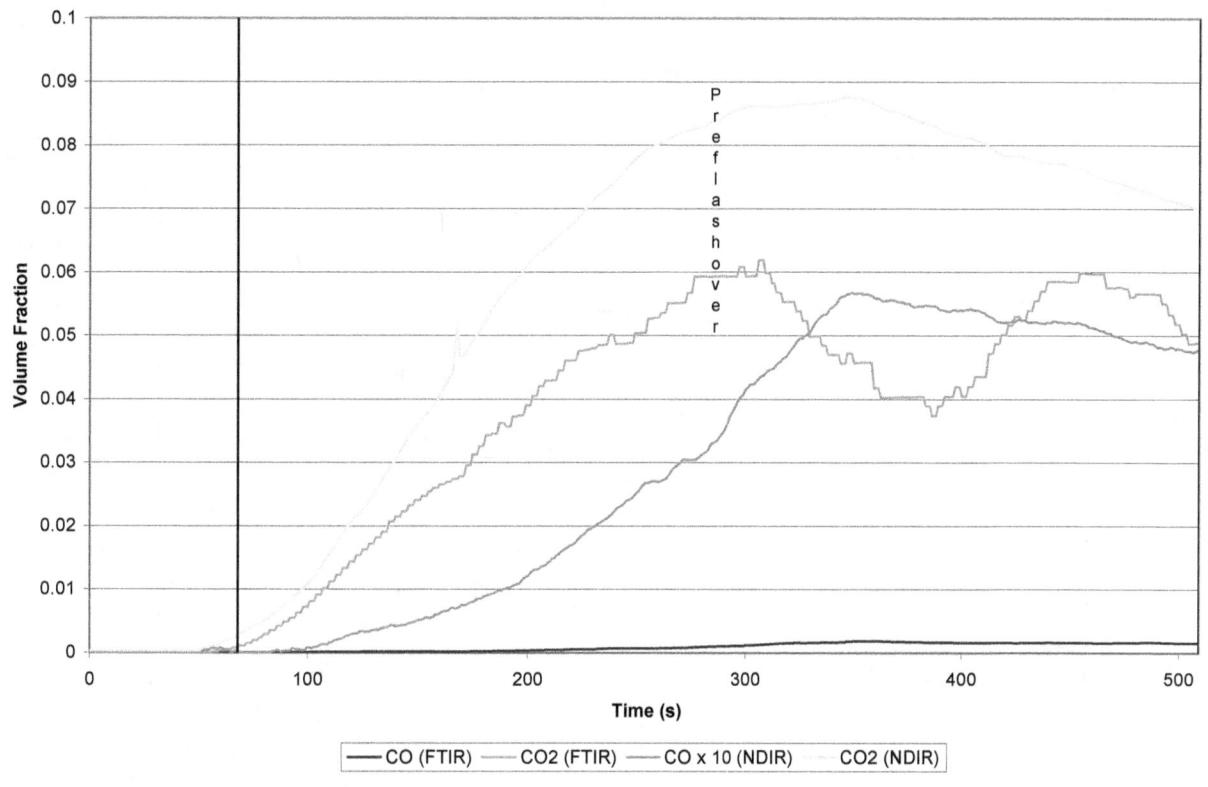

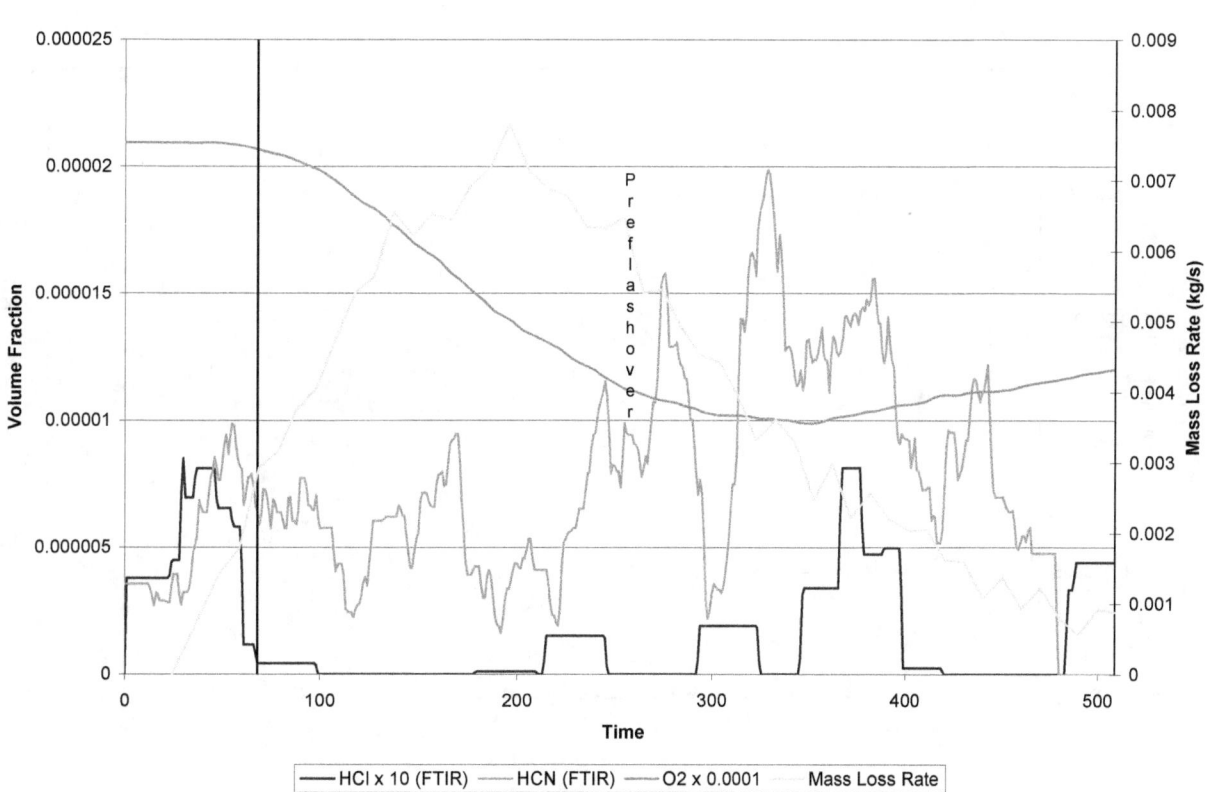

Figures A24a, A24b. Data from Test SC2, Location 1

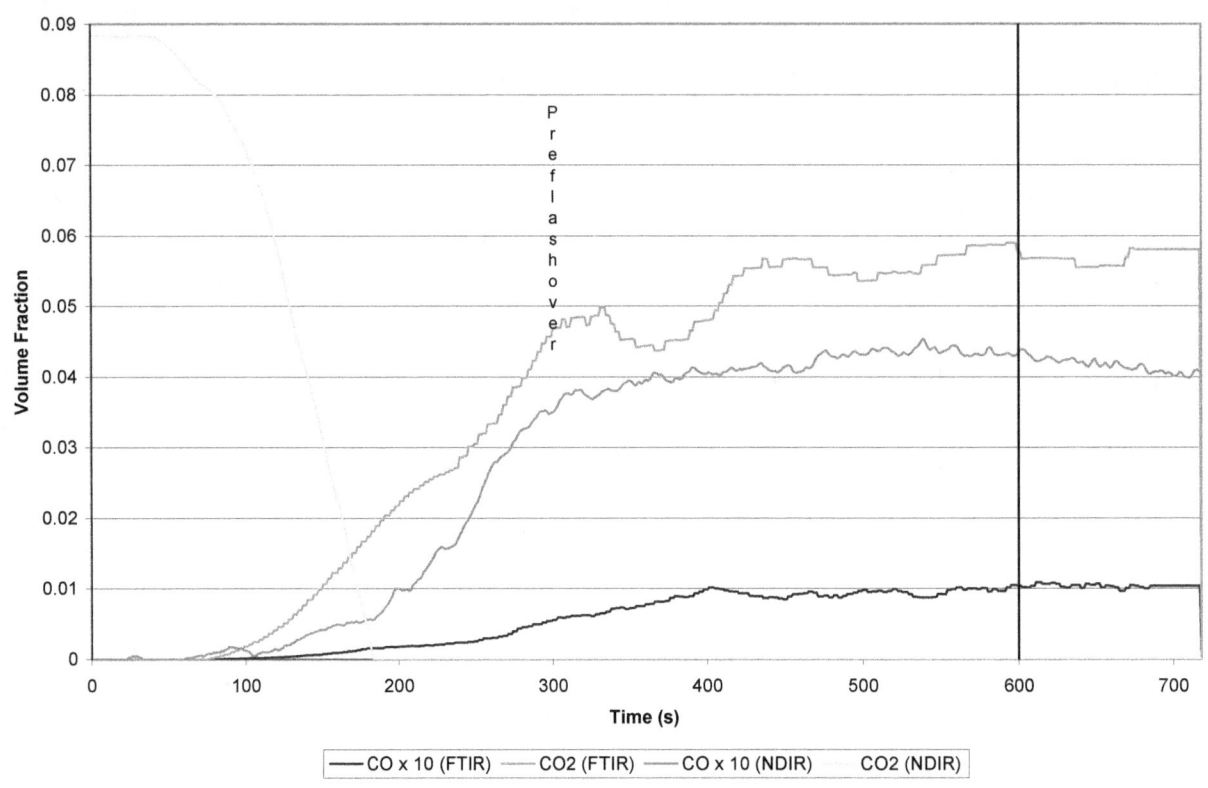

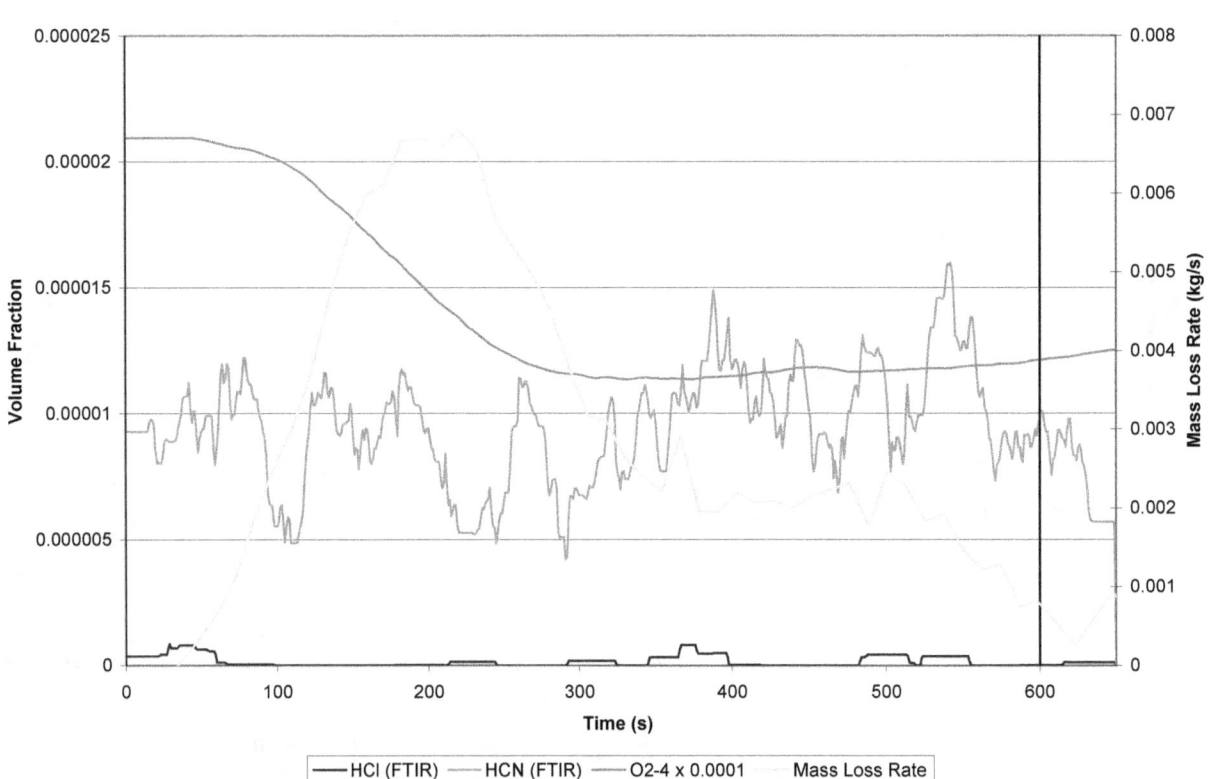

REFERENCES

[1] Phillips, W.G.B., Beller, D.K., and Fahy, R.F., "Computer Simulation for Fire Protection Engineering," Chapter 5-9 in *SFPE Handbook of Fire Protection Engineering*, 3rd Edition, P.J. DiNenno *et al.*, eds., NFPA International, Quincy, MA, 2002.

[2] Jones, W.W., "A Multicompartment Model for the Spread of Fire, Smoke and Toxic Gases," *Fire Safety Journal* **9**, 55-79 (1985).

[3] Peacock, R.D., Jones, W.W., and Bukowski, R.W., "Verification of a Model of Fire and Smoke Transport," *Fire Safety* Journal **21**, 89-129 (1993).

[4] "Standard Test Method for Heat and Visible Smoke Release Rates for Materials and Products Using an Oxygen Consumption Calorimeter," ASTM E1354-99, ASTM International, West Conshohocken, PA, 2001.

[5] See, *e.g.*, "Standard Test Method for Fire Testing of Mattresses," ASTM E1590-01, ASTM International, West Conshohocken, PA, 2001.

[6] "Life-threatening Components of Fire – Guidelines for the Estimation of Time Available for Escape Using Fire Data," ISO/TS 13571, ISO, Geneva, 2002.

[7] Gann, R.G., Averill, J.D., Butler, K, Jones, W.W., Mulholland, G.W., Neviaser, J.L., Ohlemiller, T.J., Peacock, R.D., Reneke, P.A., and Hall, Jr., J.R. "International Study of the Sublethal Effects of Fire Smoke on Survivability and Health (SEFS): Phase I Final Report," NIST Technical Note 1439, National Institute of Standards and Technology, Gaithersburg, MD, 185 pages, 2001.

[8] "Standard Test Method for Developing Toxic Potency Data for Use in Fire Hazard Modeling," NFPA 269, NFPA International, Quincy, MA, 2000.

[9] "Standard Test Method for Measuring Smoke Toxicity for Use in Fire Hazard Analysis," ASTM E1678-02, ASTM International, West Conshohocken, PA, 2002.

[10] "User's Guide to the NIST/BFRL Six Meter Hood Calorimeter," National Institute of Standards and Technology, Gaithersburg, MD, in process, 2003

[11] McGrattan, K.B., Baum, H.R., Rehm, R.G., Hamins, A., Forney, G.P., Floyd, J.E., and Hostikka, S., "Fire Dynamics Simulator (Version 2) - Technical Reference Guide," NISTIR 6783, national Institute of Standards and Technology, Gaithersburg. MD, 2001.

McGrattan, K.B., Forney, G.P., Floyd, J.E., and Hostikka, S., "Fire Dynamics Simulator (Version 2) – User's Guide," NISTIR 6784, National Institute of Standards and Technology, Gaithersburg. MD, 2001.

[12] Ohlemiller, T.J., and Villa, K., "Furniture Flammability: An Investigation of the California Bulletin 133 Test. Part II: Characterization of the Ignition Source and a Comparable Gas Burner," NISTIR 4348, National Institute of Standards and Technology, Gaithersburg, MD, 1990.

[13] Dey, M.K. "Evaluation of Fire Models for Nuclear Power Plant Applications: Cable Tray Fires – International Panel Report." NISTIR 6872 (2002).

[14] Peacock, R.D. and Babrauskas, V. "Analysis of Large-scale Fire Test Data," *Fire Safety Journal* **17**, 387-414, 1991.

[15] Pitts, W.M.; Braun, E.; Peacock, R.D.; Mitler, H.E.; Johnsson, E.L.; Reneke, P.A.; Blevins, L.G., "Temperature Uncertainties for Bare-Bead and Aspirated Thermocouple Measurements in Fire Environments." U.S./Japan Government Cooperative Program on Natural Resources (UJNR). Fire Research and Safety. 14th Joint Panel Meeting. Proceedings. May 28-June 3, 1998, Tsukuba, Japan, 240-247 pp, 1998.

[16] Newman, J.S. and Croce, P. A., *J. Fire Flamm.* **10** (1979) 326.

[17] Peacock, R.D., Bukowski, R.W., Reneke, P.A., Averill, J.D., and Markos, S.H., "Development of a Fire Hazard Assessment Method to Evaluate the Fire Safety of Passenger Trains." Proceedings of the Fire and Materials 2001 Conference, Interscience Communications, London, pp 67-78, 2001.

[18] Heskestad, G., "Bidirectional Flow Tube for Fire-Induced Vent Flows," pp. 140-145 in Large-Scale Bedroom Fire Test, July 11, 1973, P.A. Croce and H.W. Emmons, eds., FMRC Serial 21011.4, Factory Mutual Research Corp., Norwood, MA, 1974.

[19] Emmons, H.W., "Vent Flows," Chapter 2-3 in *SFPE Handbook of Fire Protection Engineering*, 3rd Edition, P.J. DiNenno *et al.*, eds., NFPA International, Quincy, MA, 2002.

[20] Link, W.T., McClatchie, E.A., Watson, D.A., and Compher, A.B. "A Fluorescent Source NDIR Carbon Monoxide Analyzer," 12th Conference on Methods in Air Pollution and Industrial Hygiene Studies, 1971.

[21] Haaland, D.M., Easterling, R.G., and Vopicka, D.A., *Applied Spectroscopy* **39**, 73-84 (1985).

[22] *Gas Phase Infrared Spectral Standards, Revision B*, Midac Corp.; Irvine, CA (1999).

[23] Speitel, L.C., DOT/FAA/AR-01/88, Federal Aviation Administration, Atlantic City, pp. 1-18, 2001.

[24] *Autoquant Users Guide*, Midac Corp.; Irvine CA, 10-15 (1995-1999). See also the internal memo entitled "Description of Classical Least Squares Analysis Technique and a Comparison of Two Correction Schemes for Non-Linearity in Measured Sample and Reference Absorbances," written by Plummer, G., 2001.

[25] Huggett, C., "Estimation of Rate of Heat Release by Means of Oxygen Consumption Measurements, Fire and Materials **4**, 61-65 (1980).

[26] Krause, R.F. and Gann, R.G., "Heat Release Rate Measurements Using Oxygen Consumption," *J. Fire and Flammability* **11** 117-130 (1980).

[27] Janssens, M. L., "Measuring Rate of Heat Release by Oxygen Consumption," *Fire Technology* **27**, 234-249 (1991).

[28] Peacock, R. D., Davis, S., and Lee, B. T., "An Experimental Data Set for the Accuracy Assessment of Room Fire Models," NBSIR 88-3752, National Bureau of Standards, Gaithersburg, MD, 1988.

[29] Babrauskas, V. & Walton, W. D., "A Simplified Characterization of Upholstered Furniture Heat Release Rates," *Fire Safety Journal* **11**, 181-192 (1986).

[30] Emmons, H. W., "Vent Flows," Section 2-5 in *SFPE Handbook of Fire Protection Engineering*, 2nd Edition, P.J. DiNenno et al, eds., NFPA International, Quincy, MA, 1995.

[31] Cooper, L.Y., Harkleroad, M., Quintiere, J.G., and Rinkinen, W.J., "An Experimental Study of Upper Hot Layer Stratification in Full-scale Multiroom Fire Scenarios, *J. Heat Transfer* **104**, 741-9 (1982).

[32] Leonard, S., Mulholland, G.W., Puri, R., and Santoro, R.J., "Generation of CO and Smoke During Underventilated Combustion," *Combustion and Flame* **98**, 20-34 (1994).

[33] Peacock, R. D., Reneke, P. A., Averill, J. A., and Gann, R. G., "Test Data, Sub-Lethal Effects of Fire Smoke," NIST Fire Research Report of Test, National Institute of Standards and Technology, Gaithersburg, MD, to be published, 2003.

[34] Levin, B. C.; Braun, E.; Navarro, M.; Paabo, M., "Further Development of the N-Gas Mathematical Model: An Approach for Predicting the Toxic Potency of Complex Combustion Mixtures," American Chemical Society. Fire and Polymers II: Materials and Tests for Hazard Prevention. National Meeting, 208th. Chapter 20. ACS Symposium Series No. 599. August 21-26, 1994, Washington, DC, American Chemical Society, Washington, DC, Nelson, G. L., Editor, 293-311 pp, 1995.

[35] "Smoke Gas Analysis by Fourier Transform Infrared Spectroscopy: The SAFIR Project," VTT Research Note, Technical Research Centre of Finland, 81 pages, 1999.

[36] Pitts, W.M., National Institute of Standards and Technology, private communication to R.G. Gann.

[37] Babrauskas, V., Harris, Jr., R.H., Braun, E., Levin, B.C., Paabo, M., and Gann, R.G., *The Role of Bench-Scale Test Data in Assessing Real-Scale Fire Toxicity*, NIST Tech Note 1284, National Institute of Standards and Technology, 1991.

www.ingramcontent.com/pod-product-compliance
Lightning Source LLC
Chambersburg PA
CBHW080251180526
45167CB00006B/2494